East of West

East of West

Cross-Cultural Performance
and the Staging of Difference

Edited by
Claire Sponsler and Xiaomei Chen

palgrave

First published 2000 by
PALGRAVE™
175 Fifth Avenue, New York, N.Y. 10010 and
Houndmills, Basingstoke, Hampshire, England RG21 6XS
Companies and representatives throughout the world.

PALGRAVE™ is the new global publishing imprint of St. Martin's
Press LLC Scholarly and Reference Division and Palgrave Publishers
Ltd (formerly Macmillan Press Ltd).

ISBN 0–312–22815–5 (hardback)

Library of Congress Cataloging-in-Publication Data
East of West : crosscultural performance and the staging of difference /
edited by Claire Sponsler and Xiaomei Chen
     p. cm.
  Includes bibliographical references and index.
  ISBN 0–312–22815–5
  1. Exoticism in literature. 2. Exoticism in motion pictures. 3. East
and West. I. Sponsler, Claire. II. Chen, Xiaomei.
PN56.E78 E18   2000
809'.93353—dc21
                                                        00–031130
                                                            CIP

A catalogue record for this book is available from the British Library.

Design by Letra Libre, Inc.

First edition: September, 2000
10 9 8 7 6 5 4 3 2 1

For Cliff,
of course

Contents

Preface

The essays in this collection are dedicated to the memory of C. Clifford Flanigan, whose untimely death in 1993 deprived us of a much-loved and admired friend, mentor, and colleague. A professor of comparative literature at Indiana University, Cliff centered his work on medieval drama specifically and cultural studies more generally. In a range of seminal essays published over the course of some twenty years, he joined an extensive knowledge of theatre history to a sophisticated engagement with contemporary critical theory. Although his focus was the Latin liturgy and early church ritual, Cliff promoted the study of culture in all of its complexity and variety. Indeed, his work was remarkable for its truly interdisciplinary and cross-cultural perspective, the only perspective, he insisted, capable of doing justice to the complex workings of culture, which refuses to contain itself within national, linguistic, or disciplinary borders.

In essays such as "Liminality, Carnival, and Social Structure: The Case of Late Medieval Biblical Drama," in *Victor Turner and the Construction of Cultural Criticism*, edited by Kathleen Ashley (Indiana University Press); "Medieval Latin Music Drama," in *The Theatre of Medieval Europe*, edited by Eckehard Simon (Harvard); "The Biblical Apocalypse and the Medieval Liturgy," in *The Apocalypse in the Middle Ages*, edited by Bernard McGinn and Richard Emmerson (Cornell); and "Comparative Literature and the Study of Medieval Drama," in *Yearbook of Comparative and General Literature*, Cliff left an indelible mark on the field of performance studies. Wide-ranging and erudite, these and other of his writings demonstrate the impressive ease with which Cliff could synthesize a vast amount of historical and textual material, finding in every case the significant cultural and theoretical issues at stake in them. Moving fluently among the various languages and literatures of western Europe, his work was notable for its ability to do justice to both the original texts and performances themselves as well as all the intervening reactions of readers, audiences, historians, and critics—a panorama of reactions that formed the reception field within which those performances have been passed on and kept alive.

Cliff's influence also manifested itself in a less tangible but no less important sphere: in his teaching, both formally within the classroom and informally as colleague and fellow scholar. In the many, many courses he taught and in his delighted conversations with colleagues, he offered an incisive model of the life of the mind. His infectious enthusiasm for the subjects he taught was balanced by the rigorous intellectual standards he embraced, and his obvious reverence for the texts he taught was matched by a gleeful delight in puncturing false pieties and self-serving pronouncements. In Cliff, the scholarly life looked like—and was—both hard work and immense fun.

In honor of that spirit of purposeful play, we dedicate this volume to Cliff Flanigan, whom we will not soon forget.

Introduction

Claire Sponsler

One of the most important claims of late-twentieth-century social theory and cultural history has been that cultures and their creations are not fixed or unchanging, but rather are constantly in flux. Within the disciplines of anthropology, literary and cultural study, theatre history, and postcolonial studies, among others, recent scholarship has begun to explore the ways in which cultures constitute themselves and are in turn constituted by constantly shifting contexts, cultural interactions, and redrawn borders.[1] Whether we are talking about people, cultures, political ideas, films, novels, songs, or plays, social groups and their cultural productions have been shown to be remarkably hard to fence in, keep at home, or "possess" once and for all. Instead, both people and their cultures circulate freely throughout the global village, cropping up in sometimes surprising places and changing shape in unexpected ways. As theory and practice have begun to recognize just how much cultures have to be captured in motion, studies of cross-cultural exchanges have become increasingly crucial, if not particularly easy to undertake. Indeed, one current preoccupation of cultural studies is with how to come to understand both the persistence and the importance of this flow of cultures, peoples, and ideas across the borders that themselves are the product of conceptual and ideological as much as geographic strategies of containment.

Looking specifically at performances that escape national boundaries, *East of West* contributes to this larger investigation of cultures in motion by taking up the question of intercultural performance as a social act, examining what happens when performances and performers are transported from their "native" lands to new locales, especially across what for much of the past 600 years has seemed the vast East-West divide. The ten essays in the collection stress the complicated ways in which performances and performers from one cultural milieu have been poached, imitated, and transformed in geographically and culturally different contexts, while looking simultaneously at how local traditions have managed to adapt to and productively transform imported cultures. Taking up issues of race, gender, and national identity in performances as diverse as a sixteenth-century French royal entry, a 1950s South African adaptation of *King Kong,* and 1980s and 90s Chinese films, *East of West* explores the complicated things that happen when performances from one geographic location are re-created within another. The crafty adaptability and resilience of cultures become evident in the diverse ways that the performances are refashioned to meet new ideological, social, and political needs.

Although the essays in *East of West* draw broadly on recent work in the field of cultural studies, the collection benefits particularly from the insights of reception theory, which provides a useful theoretical framework charting the absorption of material from one culture into another, often in ways that challenge simplistic notions of cultural dominance and co-optation. Reception theory in its various formulations offers a supple way of thinking about how appropriation can lead to social and cultural resistance as often as to submission and loss of cultural integrity or identity.[2] With its emphasis on the consumers of culture rather than its producers, on the contexts within which meaning is given to cultural material rather than on textual exegesis, and on use-value rather than determinate significance, reception theory encourages a consideration of culture from the perspective of its varied users. From the perspective of reader response, audiences are required to become active manipulators of cultural texts in the most literal sense; and in processing those texts audiences become operators, performers, and even coauthors of texts.

In particular, the notion of the resistant reader, who is able at least to some extent to refuse the dominant meanings and positionings a cultural text might be trying to impose, suggests how audiences and readers can adopt adaptive and corrective strategies of interpretation and action. Ordinarily understood, consumption is characterized by passivity and inertia, waste and absence. In the model popularized by Michel de Certeau, however, consumption is a creative activity, a way of transforming the commodities that the dominant culture assumes will be passively ingested. De Certeau defines consumption as the realm of the use of an object by those who are not its makers.[3] For de Certeau, these moments of consumption within everyday life make it possible for subjects—through various strategies and tactics—to appropriate and reshape the cultural goods and attendant ideological baggage handed to them by the producers of power. Thus de Certeau turns consumption into a form of resistance rather than a passive absorption of (mass) culture's projects. Everyday life, according to de Certeau, invents and sustains itself precisely by "*poaching* in countless ways on the property of others," deflecting the intended flow of goods and altering their meanings. In this way consumers become active participants in the processes of both production and consumption, constantly adapting the material conditions of the dominant culture to their own ends.[4]

Whatever their individual differences, the essays in the collection join in refuting the idea of a dominant Western culture pressing itself relentlessly and with a culturally homogenizing and stifling effect on non-Western cultures. Rejecting this simplistic formulation, the various contributors instead direct attention to the creativeness, resilience, and adaptability of the cultures on the receiving or co-opted end of Western—and especially American—cultural exports. Like Pico Iyer in his classic essay "Video Night in Kathmandu," the contributors to *East of West* subscribe to what might be called the magpie view of culture, seeing individuals and communities as wily thieves of others' cultural goods. Together, these essays demonstrate how cross-cultural performances are capable of creative self-preservation and reinvention, offering new ways of fashioning individual and national identities. Through acts of cross-cultural poaching, performances and their audiences are able to imagine alternate possibilities for selfhood while also negotiating anxieties about racial, gender, and national differences.

The ten essays in the collection engage the problem of East-West cross-cultural performance from a variety of perspectives and across 600 years of cultural history. The contributors examine an interdisciplinary array of texts, ranging in subject from premodern European rituals to Korean melodramas to South African and Chinese films. Taken together, these essays not only challenge traditional preconceptions about Western cultural hegemony, but also undo several other enduring presumptions about the workings of cultural appropriation. While coming to their own separate conclusions about the performances they examine, the essays all in the end concur as to both the inevitability and—perhaps more surprisingly—the utility of cross-cultural exchange.

Part I. Textual Border Crossing: Linguistic, Semiotic, Performative

In his contribution to the collection, Marvin Carlson questions the monolingual model of the Western theatrical tradition by examining the effects of foreign and pseudo-foreign passages in stage plays. Although it is commonly taken for granted that the verbal language of drama is singular, performed for a fairly homogeneous public drawn from a single community of people who "speak the same language" both in a general linguistic sense and in a more particular sense such as sharing slang and local or topical references, Carlson argues in "The Macaronic Stage" that a closer look at the

Western theatrical tradition reveals that the monolinguistic model has
never been entirely accurate. Instead foreign and pseudo-foreign elements
find their way into plays for a variety of reasons, including comic effect or
exoticism. Mock Persian (the "other" language of the Greeks) shows up in
Aristophanes' plays; scraps of Italian, French, and Spanish dot English Renaissance dramas (such as the famous "language lesson" scene in Shakespeare's *Henry V*); and contemporary plays like the linguistically dazzling
Don't Blame the Bedouins, by the Canadian dramatist René-Daniel Dubois,
mix speakers of French, English, Italian, German, Russian, and Chinese in
"an apocalyptic vision of an ultimate fusion of culture and technology."
    As Carlson reminds us, even at the most fundamental level of the spo-
ken word—the basic "stuff" of performance—dramas have always been
cross-cultural. Given that the channels of communication of the modern
world are becoming increasingly more complex, "linguistic collages" of the
sort found in macaronic theatre are likely to become more common in the
theatre. Such collages will undoubtedly raise new challenges for performers and audiences and, in particular, will require new tactics of reception.
The alien outsider who appeared as a few odd fragments in Greek theatre
has today, in Carlson's words, "become one of a fugal chorus of voices," no
single one of which can claim primacy. But together these multiple voices
offer an innovative and distinctively meaningful theatrical mélange for the
audiences of a new transnational society.
    The contributions of Claus Clüver and Karen Winstead make the point
that cross-cultural performances take place not just on the actual stage but as
"staged moments" within other kinds of cultural texts. In "Concrete Poetry
and the New Performance Arts: Intersemiotic, Intermedia, Intercultural,"
Clüver traces the development of media-crossing from the Futurist and
Dadaist sound poems of the early twentieth century, up through "happenings," and on to present-day intermedia performances such as Philip Glass'
operas or Laurie Anderson's concerts. These new performance events involve
not just a blurring of boundaries between different media but also a fusion
of performer and audience, as the spectator/listener is turned into performer
when interacting with installations, for example, or co-constructing the
"text" of holographic poetry or computer-generated virtual realities. Clüver
emphasizes the consequences these new performance events have as social
performances drawing on a wealth of cultural codes and intertexts. In
Clüver's view, a simple model of cultural transmission from one culture to
another, from performer to audience, becomes increasingly untenable in a
world of multimedia. These forms of performative text-making thus pose
important questions about "interculturality" at its broadest.
    Focusing on Concrete poetry in particular, Clüver argues that a Con-
crete poem is always border crossing in that it draws for its effect on two

or more sign systems and/or media in ways that make its visual, verbal, and performance aspects inseparable. Additionally, Concrete poetry is accessible to readers and viewers from different linguistic and cultural traditions and therefore brings together its multiple audiences in ways usually denied to other forms of poetry that depend more exclusively on verbal sign systems alone. In fact, Concrete poetry often deliberately exploits the possibilities available in, for instance, using Japanese characters to create poems in which semantic meaning (which might not be known to Westerners) is replaced by a spatial meaning (which is accessible despite linguistic barriers). Many Concrete poems also exist as performance pieces, and in these Concrete sound poems the seemingly stable boundaries between language and music are dissolved. Just as Concrete poetry pushes the limits of textuality and legibility, so Concrete sound poetry explores the possibilities of spoken sound beyond the limits of speech. In both kinds of Concrete poetry, the reader or listener is required to become an active participant in the search for meaning and even in the creation of the poem itself. In this way, Concrete poetry not only destabilizes the forms and content of traditional media but also reformulates its audience as an active accomplice in the process of artistic creation, a creation that moves freely and often quite deliberately across national and linguistic borders.

Karen Winstead's essay offers a more localized exploration of the broad theoretical concerns of Clüver's contribution. "Vulfolaic the Stylite: Orientalism and Performing Holiness in Gregory's *Histories*" points to the ways in which cultural difference can literally "play a role" in what seem to be nondramatic contexts. The essay focuses on *Histories* VIII.15–16, in which Gregory of Tours describes his meeting with Vulfolaic, deacon of a small church in northern Gaul and guardian of certain relics of Saint Martin of Tours. In Winstead's reading, this incident illustrates the performative qualities of holiness and the collision between Eastern and Western ideals. Vulfolaic was formerly a stylite—standing on a pillar day in and day out—thereby enacting in Gaul a distinctly Eastern model of holiness, given that such forms of ascetic practice were widely associated with the Orient. However, Vulfolaic gave up his pillar at the insistence of the local bishops and adopted Western practices with a vengeance. In Vulfolaic, Gregory meets a double who has mimicked the Western model so thoroughly that his career is an ironic commentary on Gregory's own.

Although Gregory's portrayal of Vulfolaic may well indicate that Eastern ascetic practices were viewed as crude and barbaric, Gregory does not simply dismiss Eastern models. Instead, Winstead argues, what Gregory criticizes is not so much the practice of imitating Oriental forms of religiosity as the practice of imitation. Thus Gregory's critique may be less an instance of Western animosity toward and rivalry with the East (although it no doubt bears traces of that) than an attack on *imitatio* itself. Throughout his writings, Gregory shows a preference for a bureaucratic, organizational Christianity over a model based on imitation of iconic holy men. In the case of Vulfolaic the two systems clash head-on. Holiness may be a form of performance, as Vulfolaic shows, but Gregory stands as an early antitheatrical critic, attempting to move spiritual practice away from the performative and toward the managerial. In Gregory, then, we see Eastern models rejected for Western ones at a crucial moment in the development of Western Christianity.

Part II. Occident Meets Orient: Nation, State, and Local Tradition

While Eastern culture offered models for the performance of various kinds of Western identities, it also served more explicitly as a resource for cultural exploration in medieval and early modern Europe. In a theoretical and historical amplification of Winstead's analysis, Kathleen Ashley considers some of the uses of exotic others in early European drama. "'Strange and Exotic': Representing the Other in Medieval and Renaissance Performance" draws on contemporary cultural studies to explore the complex representations of the "other" in European performances from the fourteenth through the sixteenth centuries. Ashley's essay examines Old Testament patriarchs, magi figures, and characters in masques who are identified in costume descriptions, stage directions, or eyewitness accounts as dressed in exotic clothing and behaving in strange ways—"the stranger the better," as one medieval manuscript says. Ashley directs our attention to "Oriental" characters in a variety of early performances, asking what cultural purposes these "strange and exotic" representational figures served.

Drawing on Steven Mullaney's analysis of the wonder-cabinet in early modern Europe, Ashley looks in particular at the role of surprise and wonder in medieval and Renaissance performances. Departing from commonly held conceptions of the ways in which medieval culture "managed" and contained otherness, Ashley asks that we consider the possibility that in a number of instances the use of exotic others on the stage was intended to offer occasions for participating in the aesthetic pleasure of wonder. Analyzing the detailed stage directions and costume lists that survive from a 1583 performance at Lucerne, Ashley argues that the Oriental patriarchs and prophets of the Lucerne performance were clearly represented as strange and exotic in ways that seemed to defy easy recuperation within dominant cultural paradigms. Ashley concludes by connecting the imaging of these Eastern "others" with cultural ideologies whose contradictions were being played out on stage and in other forms of spectacle. Her essay asks that we read these performances of exotic otherness less reductively than scholars often are inclined to do and to grant a larger role to the pleasures of wonder these performances seem to have evoked.

Like Ashley, Jinhee Kim asks that we reconsider prevailing interpretive models for understanding East-West cross-cultural performances. Specifically, Kim refutes the claim that the ready acceptance of Western models in the East eradicated indigenous traditions and ultimately the cultural identity of non-Western audiences. Her essay, "On Making Things Korean: Western Drama and Local Tradition in Yi Man-hûi's *Please Turn Out the Lights*," examines the highly popular 1992 Korean play, which has received both critical and commercial success. Although *Please Turn Out the Lights* seems "wholly Western in conception and execution," Kim nonetheless sees it as a theatrical event that is at least as Korean as it is Western. Her analysis of the play serves to reveal how Korean playwrights and theatre companies have appropriated Western drama while applying their own experience and traditional values to the resulting performances. In the end, Kim asserts, "the allegedly devastating effects of Western models are themselves a Western construction." Kim's essay thus asks us to question the supposed cultural hegemony of the West over the East.

Especially in its representation of gender through the central female characters, Pak and a woman referred to simply as "the Wife," Yi's play reveals how thoroughly Korean it at heart is. In the portrayal of these two women, the play speaks directly to the needs and anxieties of male white-collar workers, who made up a surprisingly large segment of the audience for the play. For these men, who have been caught up in a rapidly modernizing Korean society, the play offers a satisfying negotiation of the conflict between tradition and modernity, a negotiation in which the voice of the past takes precedence over Western and modern ideologies. Thus even in a play as Western as *Please Turn Out the Lights*, Kim shows, the assumption that local tradition always loses out to homogenizing and modernizing cultural imports proves wrong. Instead, indigenous culture remains surprisingly strong and resilient.

The processes of modernization and the importation of Western models into Asian cultures examined by Kim are also the subjects of Sheldon Lu's essay. Representing the Chinese Nation-State in Filmic Discourse" compares films about China and Tibet produced within the United States, mainland China, and Hong Kong. By comparing three different sites of production, Lu is able to demonstrate how much nationalism still matters for the circulation of culture. Although media theorists and cultural critics have argued that we now inhabit an age of a thoroughly global culture in which culture—and especially mass culture—circulates freely across borders, Lu's analysis shows that the heyday of the nation-state is far from over in the realm of cultural production.

Lu compares the depiction of China and Tibet in the Hollywood films *Kundun, Seven Years in Tibet, Red Corner,* and *Chinese Box* with the depiction found in mainland Chinese productions such as *The Opium War* (*Yapian zhanzheng*) and *Red River Valley* (*Hong he gu*). He then turns to recent films made in Hong Kong, such as *Comrades, Almost a Love Story* (*Tian mimi*), and *Happy Together* (*Chunguang zhaxie*). Lu's analysis reveals that despite claims about the end of nationalism, Hollywood and mainland Chinese films are locked into an ideological and political opposition between East and West that harkens back to the Cold War. In contrast, films made in Hong Kong display much greater openness to imaging cultures and identities in borderless ways. Instead of relying on rigid and stereotypical political positions, Hong Kong films explore the possibility of a thoroughly transnational ethos.

Xiaomei Chen's contribution similarly explores the dynamics of Eastern appropriation of Western theatrical models. "The Making of a Revolutionary Stage: Chinese Model Theatre and Its Western Influences" analyzes the "model theatre" (*geming yangbanxi*) promoted during the Cultural Revolution. During the first three years of the Cultural Revolution, from 1966 to 1969, when schools, libraries, and all other cultural institutions were closed in China, the purveyors of the official Maoist ideology promoted what were then known as the eight revolutionary model plays, which consisted of five Beijing operas, two modern ballets, and one symphonic work, with the last three exploring artistic forms imported from the West. At the height of the Cultural Revolution, almost everyone was compelled to see these plays for the sake, so it was said, of each citizen's political education; sometimes performances even preceded or came at the end of political meetings. Importantly, Chen notes, these performances shared a common interest in staging earlier East-West conflicts, such as the War to Resist US Aggression and Aid Korea (1950–53). By directing attention to past external conflicts, these performances masked the harsh realities of chaos and civil war within China during the peak of the Cultural Revolution. The performers of "model plays" were encouraged to adopt a Westernized style of dramatic performance to portray these moments of collective nationalist triumph in the past.

The model theatre that was promoted during the Cultural Revolution was a theatrical means of evoking the Maoist memory of a past revolution and, with its re-creation on stage, became a measure for realizing a continued revolution in post-1949 China. Importantly, although model theatre is in many ways entirely indigenous to China—a dramatic form that arose in response to specific historical and social forces during the Cultural Revolution—it is also, although much less obviously, indebted to Western traditions. Model theatre depended heavily for its success on genres, media, and techniques imported from the Occident. Surprisingly, this highly political and therefore anti-Western form of drama could not have flourished without its Western features. In fact, Occidental influences played as great a role in spurring the development of model theatre as did Chinese traditions. Finally, Chen makes clear that a remarkable feature of contemporary China's cultural scene is the extent to which theatre is political and politics has assumed an essential aspect of theatricality. Contemporary China as a nation-state is, much more than just metaphorically, a stage on which is enacted an ongoing political drama in which the state continues to play a central role.

Part III. Crossing Other Borders: The Politics of Co-optation

Shifting the focus from China to Africa, Cynthia Erb's "*King Kong* in Johannesburg: Popular Theatre and Public Protest in 1950s South Africa" investigates the intersection of performativity and race in the context of a South African musical loosely based on the RKO film *King Kong*. Her essay focuses on the jazz musical *King Kong* staged in Johannesburg in 1959. The story of a black boxer named Ezekiel Dhlamini, *King Kong* is set in Sophiatown, the legendary suburb of Johannesburg that gave birth to a thriving black popular culture and various forms of political dissent during the apartheid era. Drawing on and critiquing notions of pastiche and syncretism, Erb discusses the complicated mingling of local and imported (United States mass-cultural) influences at work in the musical *King Kong;* her larger goal is to examine the musical as a work of popular hybridity that raises important issues for global cultural studies.

Erb's analysis problematizes the relationship between local syncretic culture and mass pastiche. Most cultural critique valorizes syncretic culture for having been produced by members of oppressed groups while simultaneously devaluing pastiche for its incorporation of Western elements, valuations that are based on a privileging of "authorship" over spectatorship. Stressing reception methods instead, Erb demonstrates that even pastiche cultural works that are heavily saturated with Western mass-cultural features can be creatively appropriated by non-Western, nonelite audiences for new uses. In the case of the musical *King Kong,* however deeply it incorporated United States mass-cultural models, black South Africans who saw it brought to it their own specific experiences of apartheid and other local forms of knowledge.

In "Traveling Players: Brazilians in the Rouen Entry of 1550," Claire Sponsler considers another transatlantic movement, this one involving Native Americans, who were conscripted into European performances during the early years of the European colonization of the Americas. Her essay focuses on an elaborate spectacle staged as part of Henri II's entry into the town of Rouen in 1550. In this spectacle, a detailed re-creation of a Brazilian rain forest was set up in the French countryside, complete with parrots, marmots, and apes, as well as two Brazilian villages of fifty Tupimari imported from Brazil plus some 250 Frenchmen in disguise, who staged a two-day performance that featured scenes of industrious trading activity and a battle between two groups of the "Brazilians."

Sponsler's essay argues that the Rouen spectacle reveals a complicated act of cultural exchange, one best understood through a model of culture-as-travel, which sees cultures less as isolated villages than as teeming hotels—transient, hybrid environments where cultural encounters routinely take place. Read in the context of other sixteenth-century performances that similarly featured impersonation of New World "natives," this performance seems in part designed to ratify royal claims to an empire that transcended Europe's borders. It does so through an act of cultural poaching that steals not only the setting but also the Brazilian actors themselves, replanting them within the boundaries of Europe. But, Sponsler shows, the performance also offers the chance for all participants and spectators—Brazilians as well as Europeans—to participate in the creation of an imagined and distinctly hybrid community.

Robert Clark's "South of North: *Carmen* and French Nationalisms" brings some of the issues of Sponsler's essay into the present, as he examines the ways in which not just East-West but also South-North geographic boundaries frame both cultural appropriation and resistance leading to hybrid creations. His essay explores issues of race, gender, and national identity in the construction of the figure of Carmen. Few characters in opera are more transgressive, more marked as "other," than Carmen, who stands out by virtue of her gender, her ethnicity as a *gitana,* and her working-class status as a cigarette roller in a tobacco factory in Seville. Carmen's inherent transgressiveness has made the role the site of intense negotiations (particularly around the issues of race and gender) for producers, performers, and audiences.

Clark's essay analyzes the various stagings of *Carmen,* especially the versions starring Célestine Galli-Marié in the 1870s and Emma Calvé in the 1890s. These early realizations of the role tended to distance the character through exoticization and a concern for ethnic authenticity, while since the 1960s there has emerged a greater insistence on racial alterity, most notably in a succession of black Carmens on stage and screen, although these performers have used different approaches in realizing the role's racial markers. Finally, the essay offers a broader reflection on the role of the opera in the construction of French cultural and national identity, examining how the opera continues to play out a set of historically complex attitudes toward the exotic other.

The theoretical models for understanding and even for describing cross-cultural performance will continue to be an important subject of ongoing discussion. As rich and diverse—both historically and geographically—as the offerings in this collection are, the essays in this volume by no means exhaust the field of cross-cultural drama. For instance, the tension between official and unofficial appropriations of theatrical performances, while latent in a number of the essays, does not receive explicit attention even though there are surely important things to be learned by examining the differences between officially sponsored performances (those mounted by state or government institutions of one sort or another) and performances that take place beyond or in outright defiance of the purview of officialdom. Lu and Chen consider most explicitly the political uses of cross-cultural performances by nation-states attempting to establish a global position and to control their citizens; adding a consideration of subversive, dissenting performances that contradict these nationally sanctioned dramas would increase our knowledge of the complex interactions at work in political theatre. Surely it is not hard to imagine how adopting the material from another culture could have very different political and ideological effects depending on the relationship of the performing group to prevailing power structures.

Additionally, although gender and race are issues addressed directly or tangentially in some of the essays, a more thoroughgoing examination of the intersection of race, ethnicity, gender, and cross-cultural performance is needed. Similarly, the degree to which class must be factored in has not yet been fully considered here. Clark and Kim map how gender is implicated in the performances they discuss, Erb offers a thorough analysis of race in the South African version of *King Kong* taken up by her essay, and Sponsler considers the material and economic impact of the spectacle she examines. These suggestive analyses seem to call for fuller follow-up in other, more detailed studies.

One volume of essays does not exhaust the possibilities for future work. The essays in this collection offer themselves not as final words on the subject of cross-cultural performance but as starting points for further work. We have focused for the most part on the dynamics of East-West exchanges, since for both East and West over the past 600 or more years, those exchanges have seemed most complex, most difficult, and most daring. Broadening the geographic range or shifting the perspective to foreground other crucial cultural divides not only would bring to light other examples of cross-cultural theatrical exchanges but would also undoubtedly call for other theoretical strategies for dealing with them. We hope that *East of West* will serve as a spur to such further efforts, which will collectively add to our understanding of theatrical culture's place in a world where cross-cultural contact can no longer be ignored.

Notes

1. For a few representative studies, see Homi K. Bhabha, *The Location of Culture* (London: Routledge, 1994); James Clifford, *Routes: Travel and Translation in the Late Twentieth Century* (Cambridge: Harvard University Press, 1997); and *The Intercultural Performance Reader*, ed. Patrice Pavis (London: Routledge 1996).
2. Robert C. Holub, *Reception Theory: A Critical Introduction* (London: Methuen, 1984) provides a succinct introduction to the chief theorists and concerns of reception theory. The classic study remains Hans Robert Jauss, *Toward an Aesthetic of Reception,* trans. Timothy Bahti (Minneapolis: University of Minnesota Press, 1982).
3. A useful discussion of de Certeau's theories of consumption can be found in Mark Poster, "The Question of Agency: Michel de Certeau and the History of Consumerism," *Diacritics* 22, 2 (1992): 94–107.
4. Michel de Certeau, *The Practice of Everyday Life,* trans. Steven F. Randall (Berkeley and Los Angeles: University of California Press, 1984); the quotation is from xii. Like Bourdieu, de Certeau is aware of the way that practices tend to disappear in object-oriented analyses, especially since he sees practice as essentially a nondiscursive domain, a domain that analysis tries (often unsuccessfully) to submit to discourse. Practice, de Certeau insists, always has to be seen as ambiguous—as both beyond discourse and produced by it.

Part I

Textual Border Crossing:
Linguistic, Semiotic, Performative

Chapter 1

The Macaronic Stage

Marvin Carlson

Hieronimo: Each one of us
Must act his part in unknown languages,
That it may breed the more variety:
As you, my lord, in Latin, I in Greek.
You in Italian; and for because I know
That Bel-imperia hath practis'd French
In courtly French shall all her phrases be.

Thomas Kyd, The Spanish Tragedy

One of the first books on theatre semiotics published in English was a collection of essays by Patrice Pavis entitled *Languages of the Stage.*[1] The title referred to the observation, central to theatre semiotics, that the theatre speaks many languages, often simultaneously, not only in the dialogue but in costume, gesture, movement, setting, lighting, and so on. Rarely if ever has spoken language been the theatre's only channel of communication, and often it has not even been the most important. Much of the alternative and avant-garde theatre of the twentieth century has emphasized this multichannel aspect of theatrical art, and a very wide variety of contemporary experimental work continues this tradition. A number of major artists have emphasized visual images over the spoken text. Robert Wilson provides a particularly striking example of this, and one might also cite the entire *Tanztheater* movement in Germany, the closely related work of Maguy Marin in France and Martha Clarke in the United States, or companies like

Barcelona's Fura dels Baus, or Amsterdam's Dogtroep. Mixing film and video with live action extends the multichannel aspect of theatrical production still further, and again one might cite many important recent experimental artists and companies that have worked in this direction: Robert Lepage in Canada, Peter Greenaway in England, and in the United States most of the leading experimental groups in recent years, including the Mabou Mines, Reza Abdoh's company, the Wooster Group, and most recently the Builder's Company.

Some of these groups and individuals have kept verbal language at the center of their work, while others have distinctly subordinated it to other channels of communication, but whatever role verbal language has played, it has been normally taken for granted that this language is singular, and essentially congruent (so far as communication is concerned) with the language of the audience gathered to watch the production. This in turn reflects theatre's highly social method of creation, almost always designed and performed for a concentrated public, drawn from a single community and often even from a rather restricted subset of that community, socially and culturally—people who not only speak the same language in a general sense but also in a much more particular sense, including local slang and local and topical references. Our usual sense of this process is that the original play speaks essentially the same language as its audience, but when we come to consider cross-cultural performance, as dramatic material moves across temporal or geographic boundaries and audiences change, adjustments must be made, and among the most serious and interesting of these are the adjustments that must be made in terms of the language or languages of the performance.

The model of a monolinguistic congruence between play and audience, requiring translation into a parallel language when the target audience changes, is so familiar that it might appear almost universal, but in fact nearly every period of theatre history offers examples of plays that utilize more than one language, and our own era is particularly rich in the number and variety of multilanguage performances. Such plays might be called "macaronic," a term first coined to characterize Renaissance texts that mixed Latin with vernacular languages, but later used for any text employing more than one language. Clearly such texts offer a challenge to the communication model of semiotics and reception and raise such questions as what purposes, especially what communicative purposes, can be or have been served by creating macaronic texts and what reception strategies could be involved when an artist writing for his own linguistic community decides to include in his text passages from another language entirely.

Every macaronic performance may be seen as a cross-cultural activity, a staging of difference, although the motives for such activity have been different in different historical periods. In this necessarily brief survey of this phenomenon, I hope to show how the two most common traditional motivations for such performances, verisimilitude and the desire to appeal to linguistically mixed target audiences, have been augmented in more recent times by an awareness of the power of multilanguage productions to emphasize a variety of social, political, and cultural concerns. As the totalizing artistic theories of modernism have given way to the multiple perspectives of postmodernism, as cultural difference and diversity have gained greater interest in the world's artistic community (an interest reflected in the present collection of essays), these changes have also provided a greater interest in and impetus for productions that stage cultural difference in one of its most marked ways, through the languages spoken by the performers.

In the tradition of the Western theatre, the most common stimulus for macaronic production, until very recent times, was artistic verisimilitude. Whenever dramatists have been interested in pursuing artistic reality, a potential tension has existed, still not wholly resolved today, in the presentation of foreign-speaking characters, the tension between the aesthetics of the slice of life and the demands of artistic convention. Victor Hugo, who was quick to defend verisimilitude and truth to nature when it suited his purpose, also ridiculed theorists who demanded absolute reality on the stage, pointing out that a certain conventionality was always present. His primary example is a linguistic one: Corneille's *Le Cid*, whose hero speaks in verse and, worse still, in French, rather than in his actual language, Spanish.[2] Clearly Hugo expected the appearance of a real Spanish-speaking character on the French stage to be taken as manifestly absurd, but he did not consider that characteristic of theatre pointed out by Bert States, its constant conversion of the raw material of life into theatre, of consuming "the real in its most real forms."[3] The ancient French custom of speaking in verse in serious drama did not in fact long survive after the romantic period. Indeed the romantics, with their appeals to verisimilitude, significantly prepared the way for the development of prose dialogue, the speech of everyday life. Having a Spanish hero speak Spanish, even in a French play, obviously presented problems, but the forces of verisimilitude continued to press in that direction. In the film, a medium that carried the nineteenth-century theatre's love of realistic detail to a far greater extreme, it has become more and more common in the twentieth century to portray characters speaking languages other than the presumed language of the audience, languages that are then translated by subtitles. This practice has generally replaced the English spoken with appropriate foreign accents that I recall from films of my youth and has probably resulted from an increased audience acquaintance with and acceptance of subtitles in actual foreign films. In operatic production, the use of sub- or supertitles for this purpose has now become routine, and it is becoming increasingly common in theatrical productions that, like certain films, include characters speaking different languages. While preparing this essay for publication in December of 1999, I attended the most recent production by Robert Lepage, *Geometry of Miracles*, which utilized supertitles to translate the several scenes spoken entirely in French.

No such technological bridge between communication and linguistic authenticity was available to the pioneer stage realists of the nineteenth century, but that did not prevent them from taking on this problem. Even before Hugo, the pioneering melodrama author Pixérécourt, who prided himself on consulting *memoires* and accounts of battles to guarantee the facts in his historical spectacles, utilized a recently published dictionary of "Caribbean" language to write the speeches for the natives appearing in his *Christoph Colomb*. Fortunately for the ease of communication, the realists of the later nineteenth century favored local subjects, so the language question rarely arose, but when it did, the aesthetic of realism favored authenticity, even over communication, as may be seen in the frequent complaints of the leading French critic Francisque Sarcey that he could no longer hear what Antoine was saying when the actor insisted upon playing entire scenes with his back to the audience.

Bert States's astute observation about the theatre's drive to absorb external reality is not of course directed at the realist movement in particular but speaks to a general characteristic of the art form. Since multiple languages have been a part of the human experience at least as long as the drama, we should not be surprised to find a long tradition of theatrical attempts to negotiate with, if not absorb, this particular phenomenon. Indeed, had Hugo been forced to take into account dramatic traditions outside the traditional French serious theatre, he might have noted that from time to time dramatists did in fact create foreign characters who, in the interests of verisimilitude, spoke their own language. Not surprisingly, the two most accessible examples of this were traditions clearly operating outside of the totalizing aesthetic of the neoclassic tradition—that of the English Renaissance theatre and that of the Italian commedia dell'arte.

Although the popular impression of the commedia emphasizes its nonverbal aspects, it was in fact the most linguistically varied theatre of early modern times. In its early stages it assembled stock characters from various regions of Italy, defining them in part by distinctive accents, and this linguistic play soon expanded to include passages in other languages as well. Here as in other Renaissance theatre forms, the stock figure of the doctor or pedant invariably salted his pompous speeches with Latin or mock Latin (a trait carried on in the familiar doctors of Molière's comedies), and the ubiquitous Spanish captain often did in fact speak at least partly in Spanish, so that many commedia companies required a knowledge of Spanish for this part. Linguistic humor was always an important part of commedia amusement. There was even a traditional *lazzo* of learning to speak French, with comic and often obscene misunderstandings, a device taken over by Shakespeare in his famous "learning to speak English" scene in *Henry V*.[4]

Shakespeare and the Elizabethans, not surprisingly, offer the most elaborate examples of macaronic theatre, precisely the sort of triumph of realism over neoclassic purity and clarity that one might expect of the more open theatrical culture of Renaissance England. The already mentioned comic scene between the French queen Katherine and her servant Alice in *Henry V* is particularly striking, but a host of other macaronic scenes could be added to it from the work of Shakespeare and his contemporaries. In this same play Katherine speaks entirely in French in other scenes, and Pistol also has an extended dialogue with a French-speaking soldier. In *Henry IV, Part I,* both Owen Glendower and his daughter, Mortimer's wife, speak in Welsh, though Shakespeare here merely indicates where the lines appear and does not write them out, as he does the French passages. Shakespeare's contemporary Thomas Dekker was particularly enamored of linguistic mixing, with substantial non-English passages in six of his plays, spoken by four Dutchmen, four Welshmen, and one Irishman. The Welsh scenes in his *Patient Grissil* are so detailed and accurate that it has been suggested that Dekker may have here utilized a Welsh-speaking collaborator.[5]

Actually the incorporation of alien linguistic material into dramatic texts can be traced to classic practice. The Greek word for barbarian, *barbaros,* literally meant a non-Greek speaker, and it is not surprising that when barbarians occasionally found their way into the plays, they often brought this characteristic with them. So the Triballian god in Aristophanes' *The Birds* speaks in an incomprehensible corrupt Greek that Heracles interprets according to his own interests, the Scythian archer speaks in a similar tongue in *Themophoriazusae,* and Pseudartabas in *The Archarnians* speaks in a mock Persian.[6] Persian was the "other" language best known to the Greeks, and it also adds a touch of exoticism or attempted verisimilitude to the more serious theatre, as in the mourning laments of the chorus in Aeschylus' *The Persians.*

When we compare these Greek usages with those of Shakespeare and Dekker in reception terms, however, there is a significant difference. The use of non-Greek in Greek plays is largely ornamental or symbolic, designed only to sound generally alien or generally Persian. It is not an actual fragment of another language. The mock Turkish in Molière's *Bourgeois Gentilhomme* provides a more modern example of the same device. In Shakespeare and Dekker, real foreign speech is used, with real content that contributes, albeit never centrally, to the content of the play. It would be dangerous to assert that audience members with no knowledge whatever of French or Dutch would find these scenes meaningless—the ingenuity of actors and the multiplicity of theatre's communication channels would normally prevent that—but the full enjoyment of the scenes, and of the jokes, does require such knowledge. Shakespeare could reasonably rely on at least a rudimentary knowledge of French among his more well-educated patrons, and Dekker on some knowledge of Dutch among the not inconsiderable part of the Elizabethan public engaged in shipping and trade. What seems to be operating here is a theatre that simultaneously addresses itself to a number of different publics within a single play, a feature that has long been acknowledged as typical of the variegated Elizabethan drama, much to the discomfort of neoclassic critics.

A certain amount of the problem that macaronic theatre presents to reception analysis disappears when we recognize that in many periods, audiences were themselves macaronic, at least in part. Certainly, to return to the original meaning of this term, vernacular texts in the late Middle Ages and early Renaissance that included Latin passages could obviously assume that they would be accessible at least to the learned among their audiences. In more recent times the cultural dominance of France in the eighteenth and nineteenth centuries meant that such disparate dramatists as Goldoni in Italy, Holberg in Denmark, and Chekhov in Russia could include phrases or even short scenes in French with the apparent expectation that a reasonable percentage of their audiences would find this language quite accessible. Today this is becoming increasingly true of English, and I am often struck, especially in Germany, the Netherlands, and Scandinavia, by how often English phrases, or even speeches of considerable length, appear in contemporary productions, with no apparent communication problem, especially on the part of younger audiences.

Americans, partly due to the size and relative geographic isolation of the United States and partly due to the worldwide dominance of English, are often unaware how much of the rest of the world is in fact multilingual. In such a world, what is surprising is not that macaronic theatre appears from time to time but rather that there is not more of it. The reason for this lies in large part, I think, in the very close historical relationship between theatre and the development of modern national and linguistic self-consciousness. During the nineteenth century, as modern national consciousness developed, it was often closely tied to the celebration of a national language, which in turn was often connected to the development of a new national theatre—thus a Czech-language theatre was established in Prague as a national alternative to the earlier German-language theatre, a Norwegian-language theatre in

Christiana (today Oslo) as an alternative to the earlier Danish-language theatre, a Gaelic-language theatre in Dublin as an alternative to the earlier English-language theatre, and so on. Conversely linguistic minorities in alien lands often founded theatres speaking their own language to give them a cultural and social center, as we can see for example in the many ethnic theatres in nineteenth-century America.

These forces of linguistic, national, and ethnic identity remain strong today and probably provide a greater barrier to macaronic theatre than the more obvious problem of audience comprehension. Even in urban centers where many inhabitants speak, or at least somewhat understand, more than one language, individual theatres traditionally cater to a single language as they do to a single type of drama, such as boulevard comedy, musicals, or experimental theatre. However, in certain strongly bilingual communities, such as in the French- and English-speaking theatre centers of eastern Canada, a significant body of bilingual drama has recently appeared. Perhaps the best-known example is David Fennario's 1979 play *Balconville*, which deals with two neighboring families, one primarily French-speaking, the other primarily English-speaking, and the relationships between them and to the larger political system. It was performed with great success in Montreal, Toronto, and Ottawa and won the Chalmers Award for best Canadian play of the year.[7] As an officially bilingual city, Montreal seems a particularly appropriate location for this contemporary use of multiple languages in the service of verisimilitude, but in today's constantly shifting linguistic communities, almost any urban center could provide dramatic encounters of this sort. One of the first productions, for example, of the experimental group Stoka in Bratislava, Slovakia, the 1991 *Sentimental Journal,* concerned a young couple trying to understand each other, he speaking English, she Slovak.

Nor is verisimilitude the sole or in many cases even the predominant motive for such experimentation. Almost inevitably in any community with competing languages, this competition involves a host of other political, social, economic, and class tensions as well, and the dramatist can employ various languages as a powerful symbolic shorthand for these matters. This is what Fennario does in *Balconville,* what New Zealand author Hone Tuwhare does with his mixture of Maori and English in his *On Ilkla Moor B'aht'at* (set on the North Island, where Maori is commonly spoken, but with visiting South Islanders who speak only English), and what Brian Friel does in his *Translations,* dealing with an Irish teacher of Latin and speaker of Gaelic who is confronted with the literal and figurative encroachment of English into his domain. Certainly for Fennario's play, and to an extent for Tuwhare's and Friel's, one can postulate an audience like that I have postulated for Shakespeare and Dekker, in which at least some members can follow both languages and the rest are willing enough to accept the convention to allow all three plays a considerable popular success.

The relationship of theatrical production to linguistic heterogeneity has become a prominent concern in postcolonial theatre, where contemporary playwrights must deal with both the communication and the political problems represented in many cases by a still dominant foreign colonial language and a variety of competing native languages. Christopher Balme's recent study of postcolonial theatre devotes an entire chapter to the subject "Language and the Post-Colonial Stage." A major section of this chapter is concerned with experiments in "polyglot" theatre in such varied postcolonial locales as Nigeria and South Africa. In each of these African nations, the dramatist is confronted with a choice between utilizing the generally understood but externally imposed colonial language or selecting among a variety of competing and often mutually unintelligible local languages. Certain dramatists, however, like Nigeria's Ola Rotimi, have recognized that polylingualism is both "an ethnic fact and a political issue, the artistic and literary reflection of which can be most effectively dealt with in the theatre," and have created plays like Rotimi's 1985 *Hopes of the Living Dead,* which employs a variety of languages as an integral part of the political message of the play.[8] Similarly plays in the township theatres of South Africa, where more than a dozen distinct languages are available to the dramatist, have used multilingual theatre as a strategy "to include or exclude sections of the audience, make statements about the linguistic nature of the power structures in South African society, and reflect generally on the place of language in a colonial situation."[9]

Balme suggests that the basic message of Rotimi's play, which, he claims, uses more than fifteen different languages, is "that linguistic diversity does not automatically exclude political unity and cooperation."[10] Interestingly, another multilinguistic play has offered a similar message, on a more global level, in the home of the former great colonial power, England, whose linguistic hegemony still dominates the postcolonial world. One of the central concerns of David Edgar's *Pentecost,* presented at London's Young Vic in 1995 and subsequently at the Yale Repertory Theatre, was the difficulty of contemporary global communication, strikingly illustrated with actors speaking a variety of languages, among them English, Arabic, Russian, Turkish, Polish, and Sinhalese. Even in an international city like modern London, there was obviously almost no chance that any audience would contain spectators who could understand even the majority of the linguistic potpourri, and the immediate reaction of audiences was to be quite taken aback by the linguistic barrage, a good deal of it incomprehensible. Most of the characters in the play, however, are really no better equipped than the audience to understand the others, and as they seek for ways to communicate with each other, the audience itself becomes engaged in this process. Each character finds his or her own way into the play's discourse with whatever language skills are available, and each spectator learns to do the same. The play thus provides a powerful metaphor for a process that is being played out today not only on the national level in all parts of the world, from New Zealand to Nigeria to Canada, but also in the international world of the modern global village, where traditionally isolated language communities are increasingly confronted with the need to interact more directly and to find ways of understanding each other. A much more surrealistic and oneiric exploration of similar concerns is offered by the linguistically dazzling *Don't Blame the Bedouins,* by Canadian dramatist René-Daniel Dubois, which mixes speakers of various dialects of French, English, Italian, German, Russian, and Chinese in an apocalyptic vision of an ultimate fusion of culture and technology.

These very different plays by Edgar and Dubois are in some ways extreme cases, but the amount of multilanguage theatre being produced around the world today is considerable, and seems to be growing, just as exposure to a variety of languages is everywhere increasing. This is due partly to the growing ease of international travel, partly to the growing accessibility of film and video from many sources, partly to a new awareness of and celebration of minority languages in cultures where they were previously ignored or looked down upon by majority-language communities, and partly (as Edgar's play emphasizes) to such often less benign manifestations as the ongoing economic and political dislocation of language populations. A reflection of this latter process clearly marks the multiple-language productions of the great experimental director Tadeusz Kantor, who evokes the linguistic mélange of Poland in his youth by including speeches in Polish, Yiddish, Hebrew, and German in *The Dead Class* and in Polish, French, German, and Russian in *Today is My Birthday.*

Despite the considerable variety in cultural context and in apparent intended effect of the macaronic theatre so far discussed, all of these examples, from classic to contemporary times, share one very important characteristic. They all involve dramatic texts created, for whatever purpose, from linguistic elements borrowed across cultures. A quite different and rather more surprising sort of macaronic theatre has, however, emerged in recent years, productions in which a text originally written in a single language is given a production in which more than one language is utilized.

One widespread, if modest, example of this sort of activity may be found in tracing the history and practice of international touring. Surely, as soon as touring companies began to travel outside their own language communities they discovered the power of mixing even a few foreign words or phrases into their offerings. It is easy to see how this sort of adjustment would be quite natural to the loosely scripted and occasion-driven performance of groups like the commedia dell'arte companies, but the records of international touring are full of examples of the same sort of linguistic mixing occurring even in the comparatively fixed texts of the literary tradition. The great international stars of the late nineteenth century very often performed in their own original language while all or most of the supporting cast, recruited locally, spoke the language of whatever community the international star was visiting. Certain stars carried this mixing even into their own roles. The Italian actor Ernesto Rossi presented first individual speeches and eventually entire scenes in English while touring nineteenth-century America with his *King Lear,* an innovation that was cheered by enthusiastic audiences. In the early twentieth century, Vakhtangov's Russian company performed sequences in Polish when touring their famous *Turandot* to Warsaw. While gathering material for this essay I witnessed a current example of this ongoing practice. In a production visiting the United States from the Vaidilos Ainai Theatre of Vilnius, Lithuania, called *Mirages,* the actors had learned scattered lines in English for this occasion, and planned to keep adding lines so long as the tour continued.

In the long historical tradition of this kind of linguistic mixing, the essential goal remains the same—to establish a closer rapport with the audience, a goal distinctly more important than improving comprehension of the play. One can hardly argue that Rossi's delivery of the single line "Every inch a king" in English, his first experiment in this direction, added much to the comprehension of the play by his American audiences, but the extended cheering and applause that inevitably followed this line demonstrated clearly that the audience appreciated this as a kind of international gesture of friendship. In terms of linguistic theory, it is not so much an example of a content-bearing message as of what linguistic theorists have called "phatic communion," defined by Charles Hockett as "minimal communicative activity which has no obvious consequences save to inform all concerned that the channels are in good working order."[11]

In recent years I have noticed, however, a number of productions that make a single-language script multiple, not primarily for this traditional reason of increasing audience rapport and making a minimal communicative statement, but for the much more complicated and interesting purpose of utilizing language as one of the major symbolic structures that directly contribute to the hoped-for message or messages of the production as a whole.

One of the most fascinating experiments of this kind that has come to my attention was the production in May 1985 of Beckett's *Waiting for*

Godot by the Haifa Municipal Theatre in Israel. Haifa, like Montreal, is a binational city with deep cultural tensions and with very little cultural cooperation between the two communities. The Haifa company, sociopolitically oriented and much concerned with questions of Jewish-Israeli identity and with the Arab-Jewish conflict, has for some time included Arab actors and technicians in its company and in 1985 decided to open a new auditorium in the center of the Arab section of the city, with *Waiting for Godot,* directed by Ilan Ronen, as the opening production. The play was presented in two versions. In the new Arabic auditorium, for Arabic audiences, Vladimir and Estragon were played in Arabic by leading Arab members of the Haifa company. In this version, Pozzo spoke Hebrew to them and a distorted, limited Arabic to Lucky, while Lucky's monologue was in an accented Arabic, with some of its phrases suggesting the classic Arabic of the Koran. In the version presented to Haifa's Jewish audiences, Vladimir and Estragon spoke Hebrew with a marked Arabic accent. In both cases, the symbolic use of the macaronic production was clear, and provoked much uproar and controversy. In the words of one reviewer:

> Vladimir and Estragon appear to be two Arab construction workers of the group who come daily from the occupied territories and wait at the "slave markets" on the outskirts of Israeli cities for somebody to hire them for a one-day job. When Master Pozzo passes their way . . . he speaks Hebrew, the masters' language.[12]

For this reviewer, at least, it seems clear that the indexical use of language to encourage a current political reading of the play proved successful. Somewhat surprisingly, however, an actual audience analysis of this production found this reaction to be quite uncommon.

In this analysis, some 300 audience members from both theatres were questioned about their interpretation of this innovative production. Jewish audience members, who were more familiar with *Godot* before seeing this production, not surprisingly tended to view it within a larger context, as presenting an existential/universal problem. Rather more surprisingly, however, Arab audiences, for most of whom the play was unfamiliar, also reported seeing it in general terms. Some saw it as a general social parable, with Pozzo representing any conqueror or oppressive force; others viewed it in class terms, with Vladimir and Estragon as exploited workers and Pozzo as a bourgeois capitalist. Almost none spoke of its possible specific local relevance.

Shoshana Weitz, from whose excellent article I have gained most of my information about this production and the subsequent audience analysis, concludes from the latter a semiotic problem, a "wide gap between the

sender's intentions and what the spectators . . . received," that the majority
of those surveyed "did not respond to the rhetorical devices," including
languages used, "meant to guide them towards a local political interpretation of the play."[13] Although the responses apparently support this conclusion, I am not entirely convinced. In the highly charged cultural climate of
today's Israel, it seems to me equally likely that many audience members,
especially Arab ones, might be reluctant to admit to seeing too clear a reference to their current problems, especially if asked, as I assume they were,
by an Israeli interviewer.

In any case, in a world where for a variety of political, social, and economic reasons almost any major city now possesses sizeable populations
speaking different languages, we can no longer think of bilingual cities
like Haifa or Montreal, or indeed multilingual cities, as being exceptional.
Since the social tensions and concerns aroused by this development can
be easily and powerfully signified in the theatre by the use of different
languages, I think it is inevitable that we will see many more experiments
in macaronic theatre created specifically to comment in theatrical terms
upon these tensions, whatever the difficulties this may present to conventional ideas of reception. In 1996, for example, I witnessed in Mulheim
an experiment with certain echoes of the Haifa *Godot.* There Roberto
Ciulli created a fascinating production of Brecht's *In the Jungle of Cities*
with a cast drawn partly from Ciulli's own company and partly from a sis-
ter company in Turkey. The primary goal, I assume, was to develop international cooperation in theatre production. However, the particular
choice of Turkish, a language commonly spoken by the immigrant workers who form a significant minority of the population of Mulheim, and
the further choice to cast the Turkish-speaking actors in the "Oriental"
roles in Brecht's play, a play that is driven by economic, class, and social
conflict between Oriental and Occidental characters, seemed to me inevitably to place the use of language, as in Haifa, among the production's
political referents.

Macaronic productions of this kind, strange as they may seem at first,
can also be viewed as a recent variation upon a long-standing practice of
changing visual referents in the work of a classic dramatist like Shakespeare
to reflect local concerns. So Romeo and Juliet, for example, have been re-
cast as representing every sort of feuding family, and one could readily
imagine them today appearing (if indeed they have not already done so)
in a Haifa production, one speaking Arabic, the other Hebrew.

A somewhat more surprising, but in my experience even more common, type of macaronic experimentation is today occurring in many
places, at least in Europe and America. This experimentation does not use
conflicting languages to emphasize actual social or cultural conflict in the

original text, but rather to explore, in theatrical and performative terms, the implications of living in an increasingly multilinguistic culture. These productions seem to have developed from the recognition that in modern international culture, not only different languages but also different language speakers are being placed in contact to a greater extent than in any previous period, and that the theatre is one forum for exploring the implications of this development. Traditionally, the actors assembled to present a play, as well as their audiences, came from much the same cultural community; today both performers and audience move across cultural borders much more easily, and truly international audiences and companies are much more common than at any period in the past, vastly increasing the complexities of potential reception strategies.

Some of the most striking examples of contemporary macaronic theatre have consciously gathered into the same production actors from different linguistic and often quite different cultural and theatrical backgrounds primarily to provide an opportunity for multicultural cooperation, often with multilinguistic audiences in mind. In 1990, the Irondale Ensemble, a New York-based company that has a particular interest in the reworking of traditional texts in the light of contemporary social and political concerns, joined with the Salon Theatre of St. Petersburg to create a joint Russian-English experimental production of Chekhov's *Uncle Vanya.* Generally speaking, the St. Petersburg actors performed scenes and sequences in Russian and the Irondale actors in English, but occasional sequences or lines were also given in the "other" language in a highly complex structure that, in the Irondale tradition, mixed scenes from the original text with new material growing out of rehearsal improvisation. The goal was apparently both to develop a better understanding between artists as each language group began to experience something of the feeling of the other, and to develop a production that would provide access to both Russian- and English-speaking audiences, since it was presented in both the United States and Russia.

The London-based Théâtre de Complicité (the name itself suggests a multicultural orientation) in its *The Three Lives of Lucie Cabrol* allows the actors, who come from a variety of linguistic backgrounds, to speak a substantial number of lines in their native speech, most of it dialects of French and German. This approach not only calls attention to the international interests of the company itself, but also to the varied origins of the artists who have come together to create this work. The group's more recent *Mnemonic,* dealing again with origins, memory, and the common humanity beneath our diversity, provides several sections in other languages, with Kostas Philippoglou, for example, presenting several sequences in Greek that deal with the Greek origins of his character.

The shift from the modernist search for a "universal" theatrical language to the postmodernist recognition of the importance and signifying potential of multiple languages in a multicultural artistic milieu has been reflected particularly strikingly in the work of Peter Brook. In the modernist late 1960s Brook was much concerned with attempts to transcend linguistic variety, giving rise to such experiments as the development of the artificial "universal" language Orghast. In search of this goal he assembled an international company that attempted to find a common and universally understood artistic expression. Now, thirty years later, his postmodern reworking of *Hamlet, Qui est la?* offers a collection of fragments performed in a variety of acting styles and a variety of languages, so that Bakary Sangé, for example, playing Hamlet, and Sotigui Kouyaté, playing the ghost, both actors from Mali, speak to each other in their native language, Bambara, instead of in the "company language," French, as in previous productions.

The linguistic complexity of the former Yugoslavia has made this a region particularly rich in such experimentation. During the 1980s Kjubisa Ristic directed a company in Ljubljana and Zagreb, KPGY, that regularly presented productions mixing the various official languages of Yugoslavia. The breakup of that country has of course diminished the number of such experiments, but it has at the same time given a particular urgency to the smaller number that continue, since the aim has shifted from a fairly complacent celebration of the multiculturalism of the Yugoslavian state to a rather desperate reaffirmation of that multiculturalism in the face of the frightful cultural conflicts that have torn asunder Yugoslavia's successor states. In the summer of 1998 I saw in Pula, Croatia, several multilanguage productions, the most striking of them being *Cezar,* directed by Branko Brezovec, one of the region's best-known experimental directors. *Cezar* combined a Slovenian play from the 1920s with a contemporary play from Macedonia and material from Brecht and Shakespeare in a production utilizing actors from Slovenia, Croatia, Macedonia, and Bosnia, sometimes speaking their own languages, sometimes each other's, and sometimes bits of English and German, in a very contemporary meditation on politics, tyranny, and current cultural tensions.

One of the most complex as well as one of the most successful recent attempts at creating a production utilizing artists from different language groups, each speaking his or her own language, was the dazzling international *Midsummer Night's Dream* staged in Düsseldorf in late 1995 by the innovative Karen Beier and subsequently offered as part of the Berlin *Theatertreffen* the following spring. Beier used fourteen actors who spoke nine different European languages with smatterings of each other's languages thrown in. What may sound like a formula for confusion and frustration in fact produced an experience of astonishing richness and clarity, a solid demonstration of the artistic and communicative viability of this sort of experiment. Beier made no attempt to impose some overarching style of her own upon this disparate company. Rather she accepted the differences in style, the certainty of communication problems among the cast and between cast and audience, and the inevitable misunderstandings and surprising and unexpected communications, and wove all of these into a production that both illuminated Shakespeare's own comedy of confusion and misunderstanding and spoke fascinatingly to today's multicultural world. Traditional and current linguistic and national rivalries added extra depth to confrontations and allowed interesting connections to be made, as when Peter Quince and Hippolytus/Oberon (played by the same actor) not only both spoke Italian, but exhibited certain similar qualities suggesting the traditional Italian maestro attempting to orchestrate the unruly company under his supervision. A central reference point for the linguistic and theatrical variety of this production could be found in the rehearsal and performance of the Rude Mechanicals, who, like the actors in the larger performance, brought to their work quite varied traditions— Stanislavsky from Russia, Brecht from Germany, the commedia from Italy, Grotowski from Poland—and sought to work these together into a performance. The audience's delight in recognizing the intercultural play of these different acting styles added enormously to the fun of the whole and also reminded us that in addition to Shakespeare and to whatever "universals" of rhythm and balance may exist in any theatrical creation, the international theatre community also shares a complex tradition of performance experience that can be cited and built upon for this sort of intercultural theatre.

Perhaps the most interesting modern macaronic works, and certainly the most untraditional and most challenging to reception, are those that mix languages (almost invariably more than two) neither for the traditional motive of verisimilitude nor for the more recent recognition of varying linguistic backgrounds within the company, but rather out of an interest in linguistic mixing for its own sake, as one might mix elements of various decorative, historical, or theatrical traditions in the sort of open, decentered experimentalism that characterizes much postmodernist art. Cultural collage is a common postmodern creative strategy, and it should not be surprising that language fragments, like other cultural fragments, have begun to be worked into theatrical collage. One of Montreal's best-known experimental groups, Carbone Quatorze, led by Gilles Mathieu, not only regularly mixes French and English in its productions, as one might expect, but also regularly includes other languages as well—Italian in the recent production *Le Dortoir*, and German in the previous *Le Rail*.

Ciulli's Turkish/German *In the Jungle of Cities* also included, mostly in asides to the audience, a variety of much less clearly motivated language fragments—some Spanish, some French, some English. Michael Laub, a Belgian who directs an experimental company in Sweden, Remote Control Productions, provides a particularly striking example of such linguistic collage. In works like his 1991 *Fast Forward/Bad Air und So*, with its macaronic title, he mixes English, German, French, Dutch, and Swedish to create a fragmented multiple text operating on several linguistic and theatrical channels simultaneously.

So-called intercultural theatre has in the past normally either taken material from one culture and appropriated it for work in another, in the manner of Mnouchkine's use of Kathakali makeup and costumes in *Les Atrides*, or woven it into a modernist totalizing unity in the name of a kind of transcultural theatre in the manner of Brook's *Mahabharata*. The more recent experiments of artists and groups like Carbone Quatorze, Ciulli, Beier, and Remote Control point in quite a different direction, one that allows material, including language, from different cultures to play against material from other cultures, without necessarily privileging any particular cultural material, even that most familiar to a presumed majority of the audience. As the communicative network of the modern world becomes more complex and more intertwined, linguistic collages of this sort are likely to become more and more common in the theatre. Unquestionably these macaronic performances present new challenges and will demand new strategies for reception, but they nevertheless reflect, as the theatre has always reflected, current cultural consciousness and concerns. The alien outsider whose voice appeared only in a few grotesque fragments in the Greek theatre has today become one of a fugal chorus of voices, in some cases with none of them claiming linguistic primacy, weaving new theatrical mixtures for the audiences of a new multicultural society.

Notes

1. Patrice Pavis, *Languages of the Stage* (New York: Performing Arts Journal Publications, 1982).
2. Victor Hugo, "Preface to Cromwell," in *European Theories of the Drama*, ed. Barrett H. Clark (New York: Crown, 1965), 369.
3. Bert O. States, *Great Reckonings in Little Rooms: On the Phenomenology of Theater* (Berkeley and Los Angeles: University of California Press, 1985), 46.
4. K. M. Lea, *Italian Popular Comedy* (New York: Russell and Russell, 1962), 125–26.
5. David Mason Greene, "The Welsh Characters in *Patient Grissil*," *Boston University Studies in English* 4 (1960): 171–80.

6. A. W. Pickard-Cambridge has identified a "barbarian" character in a vase painting of the fourth century by the mock Persian word he is speaking; see his *Dramatic Festivals of Athens,* 2d ed. (London: Oxford University Press), 217 and fig. 105.

7. David Fennario, *Balconville* (Vancouver: Talonbooks, 1980).

8. Christopher B. Balme, *Decolonizing the Stage* (Oxford: Clarendon Press, 1999), 114.

9. Ibid., 116.

10. Ibid., 114.

11. Charles F. Hockett, *A Course in Modern Linguistics* (New York: Macmillan, 1958), 585.

12. B. Evron, "Bravo!" *Yediot Aharonot* (Hebrew, trans. by Shoshana Weitz), (January 15, 1985), cited in Shoshana Weitz, "Mr Godot Will Not Come Today," in *The Play Out of Context: Transferring Plays from Culture to Culture,* ed. Hanna Scolnicov and Peter Holland (New York: Cambridge University Press, 1989), 186–98.

13. Ibid., 194.

Chapter 2

Concrete Poetry and the New Performance Arts:
Intersemiotic, Intermedia, Intercultural

Claus Clüver

A t least eleven international anthologies published between 1965
and 1970 in Germany, England, the United States, Mexico, Italy,
and the Netherlands carried the label "Concrete poetry" (in their
respective languages).[1] The most lavish, Mary Ellen Solt's *Concrete Poetry: A World View* (1968), showed samples of the work of seventy-nine poets from
seventeen European countries as well as Turkey, Brazil, Mexico, the United
States, Canada, and Japan. The last to be published was the catalogue of an
exhibition, "klankteksten / ? konkrete poëzie / visuele teksten," organized
in 1970 by Liesbeth Crommelin for the Stedelijk Museum in Amsterdam,
from where it was to go to Stuttgart and Nürnberg and then to Liverpool
and Oxford. It presented the work of 140 visual poets, including nine from
Japan, and was accompanied by a record with "Concrete Sound Poetry"
by nine poets/performers. There were apparently no poets from other Far
Eastern countries involved in this movement that had begun in the early
1950s in several European countries and Brazil, had received its name in
1956 in an act of transatlantic baptism via correspondence between Eugen
Gomringer, a Swiss-Bolivian poet living in Germany, and the "Noigandres" poets, a group of young men from São Paulo,[2] had manifested itself
in numerous exhibitions and performances, and by 1970 was considered to
have run its course.[3]

Reproduced on the pages of a book, almost all of the poems strike the
reader/viewer through their visual qualities. Usually composed of a minimal amount of verbal material, frequently just one word or even single letters by themselves or in patterns of repetition, the poems rely on the
arrangement of the material in the space of the page and on the type and

size of the letters, singly or in combination. There is a simultaneous exploitation of the semantic and sound properties of the verbal medium and of the formal and expressive qualities of the notational medium—letter types or Japanese characters (*katakana* and *kanji*). Conventional grammar and syntax have largely been replaced by a "spatial syntax," and the signifying possibilities of visual design are as important as lexical semantics. Even though they disappoint many expectations of readers of poetry and demand a reorientation of approach and interpretive procedures, these poems nevertheless rely on a number of literary conventions, just as they involve the conventions of several other sign systems. A Concrete poem is always an intersemiotic and intermedia text, i.e., a text that draws on two or more sign systems and/or media in such a way that the visual and/or musical, verbal, kinetic, or performance aspects of its signs are inseparable.[4] Depending on the familiarity with multiple codes and conventions they bring to the encounter, readers may perform on such texts interpretive maneuvers that endow the often minimal material with a surprising array of significations.[5]

Among the major motivations for the creation of this kind of poetry was its expected accessibility for readers across linguistic and cultural boundaries. These anthologies were the first ever to contain so many original texts in so many different languages without translation in the traditional sense. In fact, many of these poems, though made of words, are

Figure 2.1.   Ernst Jandl, "oeö" (1964)

```
                  e
                e e
               eee
oooooooooöööööooooooo
ooooooooööööööooooooo
oooooooöööööööoooooooo
ooooooööööööööooooooo
oooooööööööööööoooooo
ooooööööööööööööoooooo
oooöööööööööööööoooooo
ooöööööööööööööööoooooo
oöööööööööööööööööoooooo
öööööööööööööööööööoooooo
eöööööööööööööööööööoooooo
eeöööööööööööööööööööoooooo
eeeeeeeeeeeeeeeee
```

untranslatable,[6] because their essence is in their structure, which often depends on the interaction of sound and sense and on the accidents of spelling, including the number of letters involved. But for that very reason it is often sufficient to provide a lexical key, the preferred procedure. Poems constructed of individual letters are not necessarily in any particular language at all, although sound and associations will vary, depending on the language within which one chooses to place them. Diacritical marks may assign a letter to a specific language, but with just a little information, even non-German speakers will immediately recognize why the intersection of a triangle of "e's" with a square of "o's" will produce "ö's" in a text by the Austrian poet Ernst Jandl (fig. 2.1).

Likewise, the rediscovery of the pictographic potential of Japanese (Chinese) characters by Seiichi Niikuni, whose work is featured in several of the anthologies (fig. 2.2),[7] becomes accessible to Western readers through the gloss explaining the ideogram and its elements (e.g., the addition of three

Figure 2.2.   Seiichi Niikuni, "*kawa/su*" [river/sandbank]

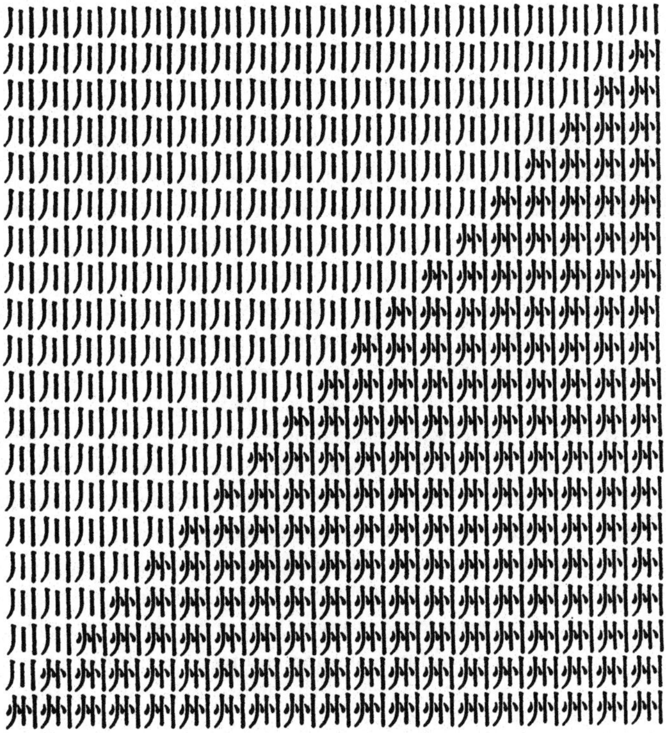

short strokes resembling grains of sand that transform *kawa*, "river," into *su*, "sandbank"), although it is difficult to assess the significance of that rediscovery within the context of Japanese writing and reading practices. Much more uncertain is the reception of Kitasono Katue's long poem "*tanchō na kūkan*" ("Monotonous Space"), printed in Emmett Williams' anthology in the original characters, in transliteration, and in Haroldo de Campos' English translation (see fig. 2.3 for Part 1 of this version).[8] Campos felt reminded of Malevich's white-on-white paintings and Josef Albers' *Homage to the Square* series, and one may also think of the poem "Sehen" from Wassily Kandinsky's *Klänge;*[9] but while these associations may make the text plausible to me, I have no idea whether this is a context within which it might have been received in Japan in 1957. On the other hand, though first published in Japan, it was composed at the invitation of Haroldo de Campos and intended for readers of avant-garde poetry in both hemispheres. John Solt, whose book on Kitasono Katue has a detailed discussion of the poem and its history, also points to the difficulty of translation: The "smooth flow" of the English version ("white square/within/white square") is achieved at the cost of inverting the mental perception, for the original has each successive square inside the preceding square.[10] Since, according to Solt, "'Monotonous Space' is poetic in Japanese because of its complex ways of inducing perceptions, not simply because of its concrete shell," this inversion may be a significant distortion.[11]

Jandl's "oeö" is about sounds but cannot be sounded, for the intersection of a triangle and a square is a visual effect. "*tanchō na kūkan*" is a poem naming colors and squares and triangles, but its effect seems to reside not so much in its visual appearance on the page as in its sound patterns.[12] Concrete poetry, usually quite inadequately defined in glossaries and handbooks, is commonly equated with visual poetry. It is true that many concrete poems defy sonorization,[13] but many others are (also) scores for oral performance by one or several voices. Each of the up to six colors that subdivide and organize the verbal material of each of the poems in Augusto de Campos' "Poetamenos" series (1953) is intended for a different male or female voice, with each color representing a separate semantic motif; visually, these motivic connections are achieved by affinity of color across the space of the implied square in which the text is inscribed. The poems were presented in performance even before they were published.[14] The fact that some Concrete "ideograms" (as the Brazilians called their texts) can be read in various directions inspired the conductor Júlio Medaglia to design a performance in which several male and female readers presented all these possibilities, passing and repassing through the text, in part in overlapping voices. Many of the artists represented in these anthologies have done significant work both as visual and as sound poets,

異 調 左 空 間

北 讃 克 衛

	shiroi shikaku	white square
	no naka	within
	no shiroi shikaku	white square
	no naka	within
	no kuroi shikaku	yellow square
	no naka	within
	no kuroi shikaku	yellow square
	no naka	within
	no kiiroi shikaku	black square
	no naka	within
	no kiiroi shikaku	black square
	no naka	within
	no shiroi shikaku	white square
	no naka	within
	no shiroi shikaku	white square

Figure 2.3.   Kitasono Katue, "tanchō na kūkan"—Part 1 (1957). With transliteration and translation by Haroldo de Campos

and some, such as Henri Chopin and Bernard Heidsieck in France, Franz Mon in Germany, the English poet Bob Cobbing, the Czech Ladislav Novák, Arrigo Lora-Totino in Italy, the Swedes Åke Hodell, Sten Hanson, and Bengt Emil Johnson, Paul de Vree in Belgium, and Jackson Mac Low in the United States, are best known for their audio-performances and/or productions. Ernst Jandl, a prolific poet and very effective performer, divided the texts in his collection *der künstliche baum* (1970), which was accompanied by a record, into "visuelle gedichte" (visual poems), "ein lesetext" (a text for reading, composed of text material found mostly on tombstones and in advertisements), "lese- und sprechgedichte" (poems for reading and speaking), "lautgedichte" (sound poems), and "ein sprechtext" (a text for speaking). His own rendition of the last, a multilingual text called "Teufelsfalle," hovers between recitation and singing, though never approaching the qualities of "Sprechgesang" invented by the

composer Arnold Schoenberg half a century before. The visual notation
of Jandl's "sound poems" seems to serve primarily as a score for his own
rendition or a memory aid for those who have heard them performed,
and the recorded performance could be considered the text proper, although no two performances would be exactly alike. The question of the
identity of the text becomes particularly intriguing in the case of recordings of performances that rely on improvisation. When sound poems or,
as Henri Chopin called them, "audio poems" exploit the ever-increasing
technologies for manipulating sound, the recording is the only text (although it may then be played as an integral part of a live performance).[15]

The notes accompanying the Stedelijk Museum's record of "Concrete
Sound Poetry" composed between 1957 and 1970 explain that the pieces
it contains range "from the single voice reading to the most complex of
machine and electronic transformations of vocal material, from the semantic and phonetic poem to poems which are a matter of vocal particles
and micro-particles rather than the word or even the letter." Just as visual
Concrete poetry has pushed the limits to various modes of illegibility so
that all the reader knows is that the "text" he sees is made up of (a) let-
ter(s), so does Concrete sound poetry explore the possibilities of speech
sound beyond the limits of speech. On the jacket of *Phonetische Poesie,* a
record he edited in the 1970s, Franz Mon distinguished with regard to
contemporary tendencies of "experimentation" between:

> two directions of exploration, one of which gropes its way along certain pa-
> rameters of spoken language—rhythm, semantics, tone color, gestures of ar-
> ticulation—and in the process uses magical singsong as well as ironic puzzles
> playing with meanings hidden in syllables (word roots) (e.g., Cobbing, Jandl,
> De Vree), whereas the other one determinedly employs technical media and
> reduces language to its microstructures, taking it beyond any familiar form
> (Chopin, Lora-Totino). The liminal experiences offered by sound poetry,
> *poésie sonore,* make us simultaneously aware of the kinds of limits along
> which language proceeds altogether: semantic, phonetic, acoustic, rhythmic,
> rhetorical. (My translation)

Mon also remarks that "electronic manipulation began to dissolve the
seemingly stable boundaries between language and music. Composers used
linguistic material, authors employed principles of musical composition."
In a number of instances, it becomes difficult to decide whether texts qual-
ify more as sound poems or as musical compositions. This is especially true
with regard to certain pieces of *musique concrète* and of electronic music,
such as Luciano Berio's *Omaggio a Joyce* (1959), which consists of electronic
manipulations of passages from *Ulysses* read in three different languages.

The Japanese composer Toru Takemitsu's *Vocalism Ai* (1960s) is basically a "sprechtext" featuring only the word *ai*, "love." Like many visual poems that allow access through a simple word gloss, *Vocalism Ai* conveys the impression that any listener can follow it once the meaning of *ai* is known; but gestures of articulation may be as culture bound as physical gestures are. We have become well aware of the misunderstandings that can be created by manual gestures or even the direction of one's glance in encounters by members of different cultures. The kind of sound poem that creates the "ironic puzzles" Mon speaks about are usually more tied to a specific language (or even dialect) and therefore less accessible to speakers of other languages than most visual poems, as exemplified by Jandl's "analysis" of Goethe's poem that begins with "Über allen Wipfeln ist Ruh," whose title, "Ein Gleiches" ("Another One Like It"), becomes highly ironic when the line appears as:

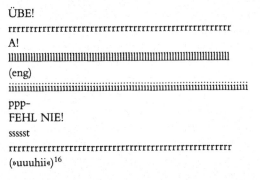

```
ÜBE!
rrrrrrrrrrrrrrrrrrrrrrrrrrrrrrrrrrrrrrrrrrrrrrrrrrrrrrrr
A!
lllllllllllllllllllllllllllllllllllllllllllllllllllllllllllllllllllllll
(eng)
iiiiiiiiiiiiiiiiiiiiiiiiiiiiiiiiiiiiiiiiiiiiiiiiiiiiiiiiiiiiiiiiiiiiiiiiiiiiiiiiiiii
ppp-
FEHL NIE!
sssst
rrrrrrrrrrrrrrrrrrrrrrrrrrrrrrrrrrrrrrrrrrrrrrrrrrrrrrrr
(»uuuhii«)[16]
```

Jandl has similarly applied this method to "translation" in his "oberflächenübersetzung" of a famous poem by Wordsworth, which renders the English sounds (without regard for lexical units) in incoherent sequences of German words that come (somewhat) close to sounding like the original. The full effect of this procedure will only be savored by readers/listeners who know both idioms and will thus appreciate the transformation of "My heart leaps up when I behold" into "mai hart lieb zapfen eibe hold" ("[Month of] May hard love cone yew charming").[17]

Most sound poets read their own texts; some collaborated with others, especially in more complex performances or productions. But there were also performers like Lily Greenham who "specialize[d] in readings of concrete and sound-poetry in 8 languages."[18] Her recording of *internationale sprachexperimente der 50/60er jahre* (ca. 1970) expects those who do not know some of the languages involved:

to listen to the texts as some kind of 'pieces of music' (sound-poetry) and to distinguish the respective colour-ranges of the languages in question. when [sic] the listener so perceives the soundmodulations of the to [sic] him [sic] foreign languages and of the human voice, he will discover that the texts project sufficient structural and phonetic fascination to rouse his interest, independent of an understanding of their semantic content.[19]

The formidable talents of this Austrian-Danish artist who has performed all over Europe make it indeed possible to experience the texts in this way, particularly when her pronunciation is most authentic. Concrete sound poems of the kind she has chosen to perform play with the possibilities inherent in the language and therefore tend to highlight its phonic qualities. But a listener who does not know Danish but understands any of the other languages represented on the record will also realize that for most of the effects achieved, the play in these texts does involve the semantics even when the material exists in a series of logically unrelated words, as in Peter Greenham's "pin point puff paper punch." He will recognize and enjoy the sound patterns of the Danish texts as performed by Greenham but find that in most instances the actual structure relies as much on semantic shifts, juxtapositions, reversals, transformations, and repetition with variations as on phonetic and rhythmic patterns. He will also realize that the tempo, rhythms, pitch level, and other aspects of the recording are Lily Greenham's decisions, the result of her understanding of structure and semantics and of the most adequate way to translate them into oral performance.

When we read that the Concrete poem communicates its own structure, a statement on which a number of the diverse attempts at definition agree and that applies to the audio as well as to the visual and the "verbivocovisual" text, we can understand it in a number of ways. As an exclusionary statement it insists that the poem does not serve to convey a message or to express a subject's feelings.[20] Concrete poems exemplify the "death of the author" that Roland Barthes was to diagnose with regard to the movement's father figure, Stéphane Mallarmé. The challenge to the reader is no longer to figure out what the poet is "trying to say" (as baffled students tend to put it) but to understand the poem's structure, or the way the poem works. The phrase that serves as the title for a collection of studies by Haroldo de Campos, "a operação do texto," is both transitive, describing the mode of the reader's engagement with the text, and intransitive: the Concrete poem operates according to an organizing principle that frequently makes it seem to be writing itself. The arrangement of Claus Bremer's highly self-referential "ein text passiert" ("a text passes by / a text happens"; fig. 2.4), which cannot be changed without destroying its effect, makes the text perform what it says. Aram Saroyan's "crickets"

```
        e
       ei
      ein
     ein
     in t
     n te
      tex
     text
     ext
     xt p
     t pa
      pas
     pass
     assi
     ssie
     sier
     iert
     ert
     rt
     t
```

Figure 2.4.   Claus Bremer, "ein text passiert" (1970)

(fig. 2.5), an iconic poem that represents the perception of the phenomenon it names even more effectively in recitation than on the page, develops according to a clear and simple principle that returns it to its beginning.[21] The same is true of a more complex text, Emmett Williams' "do you remember," which begins

> when i loved soft pink nights
> and you hated hard blue valleys

and continues by always placing first "and" then alternating between "i" and "you" followed by one of three verbs (loved, hated, kissed), of four adjectives (soft, hard, mellow, livid), of five color adjectives (pink, blue, red,

```
crickets
crickess
cricksss
cricssss
crisssss
crssssss
cssssss
ssssssss
ssssssts
sssssets
ssssкets
sssckets
ssickets
srickets
crickets
```

Figure 2.5.   Aram Saroyan, "crickets" (1965–66)

green, yellow), and of six nouns (nights, valleys, potatoes, seagulls, dew-drops, oysters): the text, which can be rendered by one or several voices, inevitably ends when the system produces again the line "and I loved soft pink nights?" Readers can (re)produce such permutational texts by following the program.

A developmental principle operates in a kinetic text by Décio Pig-natari, published as a booklet: each page adds a horizontal beam to the vertical beam seen on the first (and which only on turning the page is un-mistakably recognized as a capital letter, "I"). The progression leads in-evitably to a block-like shape formed of two verticals and three horizontals: not a part of any Western alphabet, it is recognized by the lit-erate as the Chinese character for "sun," which then explodes with the turn of the page into "LIFE" (fig. 2.6). Using the same typeface but less bold, stretching each letter to fill a square, and making the squares touch, a "logogram" by Pedro Xisto exploits the accident that in Western translit-eration "ZEN" consists of an "E" flanked by letters that in this square for-mat have an identical shape when one is turned by 90 degrees (fig. 2.7). The addition of one vertical and two horizontals results in a simple, har-monious structure with bilateral symmetry that is still legible as "ZEN" and, as a graphic design, represents what it names in accordance with the reader's understanding of the concept and the design. Inevitably, different cultural backgrounds will produce different readings. Chinese or Japanese

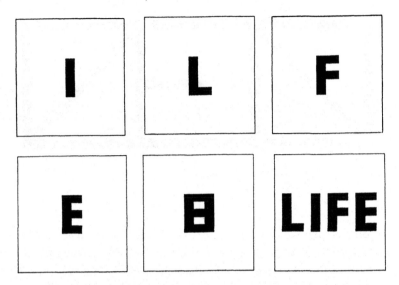

Figure 2.6.   Décio Pignatari, "LIFE" (1958)

readers may begin by focusing on the center: before ever deciphering the letters of the alphabet, they are likely to recognize the presence of the character for "sun," which may altogether escape Western readers.[22]

The readers' operation of the text is performed in several modes. They may have to handle the text as a physical object, opening, unfolding, turning, and even tearing it, or walking around it or through it; for the modest appearance of such texts on the page is only one possibility for the visual poem to exist. Pignatari's booklet "LIFE" was an insert in a portfolio edition of poster poems. (The reduction of its individual pages for the Solt anthology to the format shown in figure 2.6 spoils the process of discovery with its surprise appearance of the Chinese character.) Exhibitions require larger formats, and the demand was met not only by creating poster poems but by sculpting or building poems, painting them on canvases or sidewalks, placing them on billboards and house walls, or hanging them letter by letter from clotheslines. The performance of sound poems may require skills and demand a number of decisions not needed for the recitation of conventional lyrics. These decisions are based on another kind of operation: the active encounter with a text whose "content" is its structure, a signifying structure, to be sure, but open to multiple procedures of meaning construction, among which an exploration of the signifying potential of structure and medium in different cultural contexts may be a central concern. In an early essay Eugen Gomringer invoked the metaphor

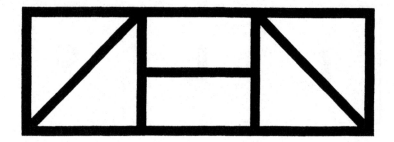

Figure 2.7. Pedro Xisto, "ZEN" (1966)

of game playing with regard to his "constellations," as he called his Concrete poems, and assigned to the poet the role of "determin[ing] the play area, the field of forces, and suggest[ing] its possibilities, the reader, the new reader, responds to the sense of play," for which a knowledge of the possibilities of the game is essential.[23] The rules of the game inhere in its structure; but the reader may be a more imaginative player than the poet who has invented them.

In 1979, Umberto Eco used a different, performance-oriented metaphor when he gave an American edition of nine of his essays the title *The Role of the Reader*. In the earliest of these, "The Poetics of the Open Work" (1959),[24] the reader's handling of certain kinds of texts is literally called a "performance" in the same sense that a pianist's rendition of a musical score is a performance. In fact, Eco also distinguished between two kinds of musical performance on the basis of the score to be performed, citing as instances of the second kind recent compositions by Karlheinz Stockhausen, Luciano Berio, Henri Pousseur, and Pierre Boulez, which require that the performer make certain decisions regarding the duration of notes or the sequence in which individually presented "note groupings" will be performed. A phrase in Pousseur's description of his *Scambi* that Eco quoted recalls Gomringer's description of the new poetry: "*Scambi* is not so much a musical composition as *a field of possibilities,* an explicit invitation to exercise choice."[25] Works like these, in which "the author offers the interpreter, the performer, the addressee a work *to be completed,*"[26] fall into the category that Eco called "open," which he contrasted with relatively "closed" works. But coming to realize that interpretive operations can also open up seemingly "closed" works, Eco made a further distinction and created the label "works in movement" for "open" works like *Scambi* "because they characteristically consist of unplanned or physically incomplete structural units."[27] Such works can be found in all contemporary media and sign systems; their

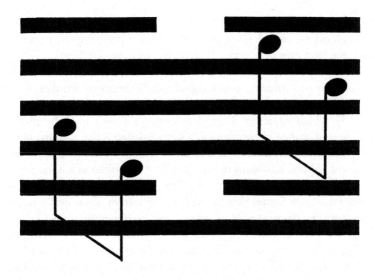

Figure 2.8.   Augusto de Campos, "Pentahexagram for John Cage" (1977)

forms reflect ways of seeing reality that Eco identified also in the sciences and modern philosophy and that he characterized as valuing ambiguity, indeterminacy, mutability, multiple polarity, discontinuity, and process.

These themes were continued a few years later by Haroldo de Campos in "A Arte no Horizonte do Provável," an essay that added "probability" to the key words just enumerated, along with such terms as "permutation," "movement," and "chance."[28] A few years before Eco's essay appeared, Haroldo had already published a short article called "A Obra de Arte Aberta,"[29] applying a term Pierre Boulez had used, to texts by Mallarmé, Joyce, Pound, and Cummings, with references to Calder's sculptures and compositions by Anton Webern, but without exploring the reader's task. In 1963, he considerably expanded the list to include the work of contemporary writers, composers, and visual artists, a number of them Brazilians, in which "probability is integrated into the very making of the work of art, as a desired element of its composition" (17). In music, he found "indeterminacy cultivated to the highest degree" in the aleatoric compositions of John Cage that were based on manipulations of the *I Ching*, the Chinese *Book of Changes*, with its sixty-four possible permutations (21). Regarding poetry, not unexpectedly, he ended up by pointing to the work of the Noigandres poets, including his own: quite a few of their texts had been composed by employing principles of permutation, and the "open matrix" of many of their poems "would permit the reading to proceed in various directions, vertically or horizontally, isolating and emphasizing [text] blocks or, instead, integrating them with other component parts of the piece by means of relations of similarity or proximity" (30).

Most of the characteristics of the kind of texts I have discussed so far, and of their performance, are exemplified by a seemingly wordless poem by Haroldo's brother Augusto, a "Pentahexagram for John Cage" (1977; fig. 2.8), which expresses their admiration for Cage's work by playing with several chance phenomena in a simple design using minimal materials. It is based on the codes and conventions of several sign systems that represent cultural traditions of East and West. Inscribed in a horizontally extended hexagram of the *I Ching* (the 49th, *Ko* / Revolution, Lake above Fire)[30] are figures that in the Western system of musical notation represent two identical pairs of descending quarter-notes placed in such a way that, if the hexagram is mentally transformed into a pentagram (and thus into a musical staff) by removing the uppermost line, they happen to spell out C A G E according to one of the conventions of naming notes. Permutation to the other implied pentagram by mentally removing the lowest line yields A F E C—which is semantically empty as a letter sequence unless the reader-operator can find a match in any lexicon or list of acronyms. Vacillating between sense and nonsense on the level of the letter values, as a musical score the text makes equal sense in either version. What the reader will do with the text beyond this point will largely depend on his "cultural lexicon," as Eco has called it, and on his familiarity with the individual "named" by the notes (and the poem's title). He may dwell on the accidents that allow the name to be represented in note language (by two descending minor thirds) and recall a similar game in music history with the letters BACH (according to a somewhat different note-naming convention), which resulted in a well-known composition. This text, however, does not seem to invite actual musical performance (although a performer accustomed to executing the visual scores of Cage and his contemporaries may find a way of making even this minimal "open" score musically interesting); instead, as a combination of visual signs it calls for silent contemplation, in keeping with Cage's fascination with silences—and with chance (playing with the letters, the reader may realize that C A G E can be filled out to create CHANGE, which the removal of a small trace will transform into CHANCE). A reader satisfied with constructing the text as an "emblematic portrait" of Cage (to use Richard Brilliant's term)[31] may be satisfied with taking the hexagram as representative of the *I Ching* and the use Cage made of it. Someone curious enough to find out why this particular hexagram was chosen may consult a translation of the *Book of Changes* and make his findings part of his interpretation, whereas some-

one raised in its traditions may approach the entire poem via the hexagram in ways only available to the initiate.

Concrete poetry lends itself well to demonstrating the possibilities inherent in intersemiotic and intermedia texts even of minimal proportions,
to exploring the potential of such texts to appeal to audiences across linguistic and cultural boundaries even when different writing systems are involved, to indicating new demands made on readers as texts leave the page
to become three-dimensional objects that require physical manipulation,
and to exemplifying new concepts of performance with regard to the text
executing itself as a programmed structure, the oralization of "open" verbivocovisual texts, and the role of the reader as game player, operator, and
quasi-coauthor.

Of course, Concrete poetry was by no means the only project that
challenged traditional modes of performance, and not everything about
it was exactly new. A number of the poets represented in the anthologies of Concrete poetry worked also in other genres. A number of them
belonged to circles or groups that engaged in public performances, "actions," and "happenings." The Stedelijk Museum catalogue contains a
long list of little magazines and periodicals from twelve European countries and Canada, the United States, Argentina, Brazil, Uruguay, and
Japan where their material was published, along with other "experimental" texts (225–27), followed by a list of sixty-six "exhibitions and
manifestations" of visual poetry that were held between 1959 and 1970,
including two in Tokyo (228–29); it is incomplete. The titles of a few of
the more important ones, several of them accompanied by extensive
catalogues, suggest the many points of contact and even transgression of
writing and the visual arts: "Schrift und Bild" (Baden-Baden, 1963),
"Between Poetry and Painting" (London, 1965), "The Arts in Fusion"
(Philadelphia, 1966), "Parola e immagine" (Firenze, 1967), "Text Buchstabe Bild" (Zürich, 1970).[32] The most interesting title is that of a much
later exhibition, held in 1993 in Marseille, for which a most lavish catalogue was produced: *Poésure et Peintrie: «d'un art, l'autre».*[33] Interestingly, the catalogue contains substantial material about sound poetry,
including a "chronologie de la poésie sonore" compiled by Dom
Sylvester Houédard (598–90). An equally interesting exhibition had
been displayed in 1980 in the Musée d'Art Moderne of Paris: its catalogue is called *Écouter par les yeux: objets et environnements sonores.*[34] It was
modeled after an exhibition held earlier that year in Berlin: "Für Augen
und Ohren." Yet another way of showing the intermedia character of
much contemporary creative production was the exhibition curated by
Dietrich Mahlow in 1987 in Mainz, again with a rich catalogue: *Auf ein
Wort! Aspekte visueller Poesie und visueller Musik.*[35]

When Hanns Sohm turned his substantial collection of objects, pictures, books, and other materials, begun in the early 1960s, over to the
Staatsgalerie Stuttgart, the new curator, Thomas Kellein, mounted an exhibition and prepared a comprehensive catalogue *«Fröhliche Wissenschaft»:
Das Archiv Sohm,*[36] which gives an excellent overview of the activities and
events that characterize the 1950s and 1960s and are documented in the
collection. Among its chapters are titles such as "From Burroughs to the
Beat Scene," "From Cage to the Happening," "FLUXUS," "Intermedia Art
in Europe and the USA," "Concrete Poetry," and "Underground and Politics." I have concentrated my investigation in this study on Concrete visual and sound poetry because it allowed me to show the new
understanding of performance that developed during that period by referring to a relatively coherent yet diversified body of texts produced in many
languages and cultures, and on some of the events surrounding and involving them. As Kellein's work suggests, the Concrete poetry movement
(as he understands it) was interconnected with a series of parallel developments, and the whole array of creative activity surveyed there is of course
only part of the entire spectrum of Western art making in the two decades
following the war—it may not even represent the mainstream.

    Yet it is possible to indicate some significant further developments of
the aspects of performance I have identified by keeping more or less close
to the kinds of texts related to those already discussed. An excellent
overview of twentieth-century visual poetry has been provided by Johanna
Drucker in "Experimental, Visual, and Concrete Poetry: A Note on Historical Context and Basic Concepts."[37] Rather than summarizing some of
her findings, I will be satisfied with pointing to an ever greater and more
inventive use of text-making practices departing from the often rigorous
principles of "classic" Concrete poetry, employing a greater wealth of materials and combining in numerous ways words and images (a practice not
very common in Concrete poems), with collage (recycling of preexisting
materials), a favorite technique in constructing such mixed-media texts.[38]
Japanese visual poets participated fully in these developments, often blend-
ing Western texts and images with Japanese or relying on the new international language of the photographic image, otherwise drawing on the
resources of their own language, script, and visual arts and their traditions,
as was beautifully documented by the exhibition "Visuelle Poesie aus
Japan," which circulated from 1997 to 1999 in Germany before being
shown at the Goethe-Institut Kansai in Kyoto/Osaka, accompanied by a
catalogue prepared by Klaus-Peter Dencker.[39] A piece like Hiroo
Kamimura's "En hommage à Gertrude Stein," composed in 1997 and reproduced in color in the catalogue,[40] appears quite accessible to a Western
reader, who will recognize against the black of the vertical rectangle a small

cut-out of Stein's head, as portrayed by Picasso, looking down at a square that contains an incomplete version of her own "Completed Portrait of Picasso," in the original on the left, which then appears to be paired with a Japanese translation typed in horizontal lines on the right, with the two texts separated in part by a pen-and-ink line drawing of a profile that seems to be a self-portrait of Kamimura. Unclear to this reader is the significance of the blue of the horizontal and the brown and orange of the two vertical color bars: if any symbolism is involved, does it rely on Western or on Japanese traditions?

When Japanese text is joined to art or objects founded on a native tradition, a mere verbal gloss may be insufficient for a Western viewer. The Dencker catalogue shows a color photograph of a 1994 piece by Chiharu Nakagawa in which a stele-like object of smooth wood sitting on a carved base has been inscribed with one character; a gloss below the photo explains that the character means "poem" or "poetry." The object, though noble, seems to be prefabricated; it is apparently the inscription that provides the surprise or friction. The title *Alibi Ware,*[41] identical in German and English, adds another metaphoric layer to the work. It helps to learn that the object is used by Japanese Buddhists to commemorate the dead, who are given a new name that is then inscribed on a stele like the one pictured; but this does not guarantee that a Western reader will assign the work a similar meaning to that produced by a Japanese or Chinese reader (if the object's cultural coding does indeed extend beyond Japan).

As we have seen, already in the Concrete phase the poem had a tendency to leave the page. It became a poster, a billboard text, was found on sidewalks or on huge inflated pillows.[42] New forms of artist's books were invented, often in order to achieve kinetic effects, as in some of Emmett Williams' projects.[43] The new technologies, already heavily exploited by sound poets, increasingly attracted visual poets as well. One of the tireless veterans is the Portuguese E. M. de Melo e Castro, who has accompanied his own work in "media poetry" and "video poetry" with theoretical, critical, and historical studies, as in *Poética dos Meios e Arte High Tech.*[44] The topical issue on *New Media Poetry: Poetic Innovation and New Technologies* of the periodical *Visible Language* (1996), edited by Eduardo Kac, has contributions on hypertexts, video poetry, virtual poetry, literary cybertexts, and holopoetry, capped by Eric Vos's essay on "New Media Poetry: Theory and Strategies."[45] A large sampling of texts of many of these types, along with manifestos and other documents, is found in the multilingual DOC(K)S 3.13/14/15/16 (1997), published in France complete with CD-ROM *Alire* 10 (May 1998); the CD *Alire* is designated as "Le Salon de Lecture Électronique."[46] Along the bottom, back, and front of the cover of

DOC(K)S runs the motto, "verslaccomplissementdeladissolutiondesiden-titésmacromoléculairesindividuellescollectivesdanslespacetemps." The movement of words and images in two- and three- dimensional space, or space-time, of computer-generated texts, involving permutations and transformations, dissolution and fusion, brings about an illusion of a self-performing text that was only suggested in the permutations and similar structural devices employed in the poems we looked at before. Theorists insist that the shift of texts to these new media and into virtual space should not be understood as merely a sophisticated extension of two-dimensional, page-bound visual or verbivocovisual poetry. Vos rightly claims that "the new media environments allow and in fact invoke the use of sign features that are very hard to employ, if at all, in a print environment" and that "many of the new media poetries . . . at least in part aim at *repletion of the verbal sign*" (232). Such aims make, as we shall see, new demands on the reader/performer, even when he is not asked to do physically much more than push a button or click a mouse.

Before reviewing the role of the reader as performer, we must note that "performance art" has in itself become a more varied and more widely practiced art form. While in the majority of computer-generated texts the presence of the artist/author has remained as effaced as in most Concrete poems, the performance tradition in which sound poetry has its place has become richer and appears to have broadened its intercultural appeal by either relying more on nonlinguistic signs or using English as the new *lingua franca*. Documentation of such performances is rarer and, on the whole, less accessible, with still photographs serving as the major and woefully insufficient source of visual information.

Though they may not in every case have been aware of the tradition when they began, the sound poets of the 1950s and 1960s soon demonstrated their familiarity with at least some of the important texts of the first half of the century, primarily those produced by Zürich or Berlin Dadaists and by Kurt Schwitters, but then also by the Italian and Russian Futurists.[47] There were some recordings, of Schwitters reciting portions of his *Sonate in Urlauten,* of Gertrude Stein reading her poems; and Raoul Hausmann recreated many of his own texts for the record. But even of Schwitters we do not seem to know exactly how he would perform his *Sonate;* I was fortunate enough to see three quite different productions of the piece see in one season (1987–88), by one or several performers, each of them intriguing, none of them claiming to reconstruct Schwitters' own performance.

There are, however, sufficient documents regarding the performances by Hugo Ball and his collaborators at the Cabaret Voltaire in Zürich to show that they involved (at times) elaborate costuming and staging, as well as a good deal of improvisation. In *Dada Performance,* Mel Gordon has collected a number of surviving performance texts ranging from Zürich via Berlin, Cologne, and Hanover to Paris and covering the years 1916 to 1924.[48] The performances and *Aktionen* in the 1950s by or involving members of the Wiener Gruppe, some of whom were also producing Concrete poetry, appear to have been entirely in the Dada spirit, knowingly or spontaneously.[49] That is likewise true of events staged by the Darmstadt Circle (with Emmett Williams and Claus Bremer among its members), by John Cage and his collaborators, and by FLUXUS (which included some of the same artists). Gordon rightly claims that "the Dadas and their work become critical in understanding . . . the phenomena of happenings, performance art, and Robert Wilson" (8).

It is difficult and to some extent pointless to extricate from this web of interconnected developments those that foreground sound and visual poetry. "Chronologies" of both types of poetry originally prepared by Dom Sylvester Houédard were revised and amplified by several hands for publication in *concerning concrete poetry*, compiled and edited by Bob Cobbing and Peter Mayer in 1978;[50] these chronologies include exhibitions, festivals, and recordings besides the production or publication dates of important texts; that more recent festivals, at least in Portugal, have mixed nonverbal performance art with the performance of "poetry" of different sorts, including "the body as medium." In mounting an exhibition of Portuguese Concrete, experimental, and visual poetry from 1959 to 1989 at the University of Bologna, the organizer of these events, Fernando Aguiar, included in the catalogue a section on "Readings, Lectures, and Poetic Performances, 1967–1988."[51] There has apparently been a lively activity in Japan involving such kinds of performances, but to my knowledge it has not yet been adequately recorded and documented for Western readers; a glimpse of it is caught in the catalogue for the 1997 Japanese Visual Poetry exhibit, which illustrates and describes Shozo Shimamoto's project of having individuals write messages on his shaved head, as his own contribution to the international "mail art" movement.[52] The most integrated intermedia/mixed-media/multimedia performance known to me is the Brazilian Arnaldo Antunes' *NOME* (*Name,*1993), published on videotape, accompanied by a book and an English version of the Portuguese texts.[53] The thirty pieces on a great variety of topics, many exploring aspects of speech and language, are performed off-screen in ever-varying modes of recitation or song while their written/typed/stenciled visual representations in many text types, usually in motion, at times transformed and otherwise manipulated, are shown against a different kind of background in every piece, either illustrating the text or contradicting it or offering a visual commentary. The former pop star has gone on tour performing the sound score live against a projection of the video.

The role of the reader, finally, as operator, cocreator, and performer has gained in complexity and importance—not coincidentally with the rise of that role in critical theory. The "text-sound texts" collected by Richard Kostelanetz in 1980,[54] which include pieces by such musical composers as John Cage, Kenneth Gaburo, and Philip Glass, for the most part challenge the reader to figure out how they might be performed; the same is true of the pieces in the "Klangbilder" section of Dietrich Mahlow's catalogue *Auf ein Wort!* (1987). In fact, in a number of cases the decisions may be whether they are to be performed at all or whether they are meant to be received "as if they were music," whatever that may mean in each case. The visual poetry of the 1960s and beyond (as Drucker and others have pointed out) shows the impact both of Pop art and of Conceptual art, the two major currents of mainstream art at that time; in fact, a good deal of Concrete poetry matches its emphasis on the materiality of the text with strong conceptual leanings before Conceptual art got its name. Affinities to (usually intermedia) Conceptual art also mark a number of the text-sound texts in the Kostelanetz book and many of the items exhibited by Mahlow: all of these are therefore open texts, and many show the tendency toward performance (and thus an emphasis on process rather than product). In his study *The Object of Performance: The American Avant-Garde since 1970* (1989), Henry M. Sayre explored a strand of postmodernist developments rooted in a modernism "that might be said to be founded in dada and futurism and that is oriented, in one way or another, to performance, or at least performance-oriented art forms" (xi); one of these he labeled "critical performance," which he exemplified by the work of Roland Barthes (so influential in the United States that he could be considered part of its avant-garde), the creative reader-performer. Where I have used the concept of openness, following Haroldo de Campos and Umberto Eco, Sayre consciously chose "undecidability," rather than "indeterminacy" as Marjorie Perloff had used it in her *Poetics of Indeterminacy: Rimbaud to Cage,* because "it seemed to locate the question of the work's contingency, multiplicity, and polyvocality in the *audience* rather than in the work itself" (xiv).[55] This is a quality that we have encountered throughout this study in works produced on four continents.

Happenings often contained invitations to audience participation, and environments always included the viewer physically and thus made him a participant, and often a performer. Allan Kaprow's *Words* (1962, repeated 1967), a space crowded with pieces of writing hanging on walls and from the ceiling, photos of which are found in Richard Kostelanetz' *Imaged Words & Worded Images* (37–39), is "a collaborative environment that the spectators continually create within a circumscribed space" (Kostelanetz). A very different kind of environment, Ian Hamilton Finlay's "Little

Sparta," the garden and temple at Stonypath, Scotland,[56] studded with inscribed sculptures and architecture, allows the visitor no more than to choose his path and thus change the sequence and perspective in which to approach each mixed-media or intermedia text, but makes great demands on his abilities as reader and interpreter. Entirely dependent on the reader as cocreator is the holographic poem (or holopoem) as created by Eduardo Kac: immaterial, it remains invisible until the viewer/reader moves into a particular position, and then the three-dimensional text unattached to any surface moves with the viewer's every movement, changing color and shape, with its letters transforming themselves into other letters or breaking up to recompose themselves into (an)other word(s), suggesting a range of meanings.[57] Engaging with interactive computer texts and hypertexts, the reader/performer has finally begun to move fully into virtual space and into possibilities of textual operations that are still being explored by the artists/programmers.

In recent years this interactivity has entered the Internet space and thus become truly global (shades of McLuhan!). On November 4, 1999 I received the following invitation via E-mail:

> Log on to: http://uirapuru.ntticc.or.jp
> UIRAPURU OVER THE AMAZON
> AWARDWINNING ARTWORK MERGES TELEPRESENCE WITH VIRTUAL REALITY, WHILE
> PINGBIRDS SING ACCORDING TO INTERNET TRAFFIC
> "UIRAPURU," EDUARDO KAC
> ICC Biennale '99, October 15 - November 28, 1999
> . . . The work is being exhibited both online and in a gallery at the InterCommunication Center (ICC), Tokyo. . . . The multi-user VRML world and the live streaming video and audio of "Uirapuru" can be experienced at: http://uirapuru.ntticc.or.jp. Remote participants are encouraged to navigate and interact with "Uirapuru," a colorful, flying telerobotic fish which hovers above a forest at the exhibition site. Documentation of the project is archived at: http://www.ekac.org/uirapuru.html. "Uirapuru" is the name of an actual bird and also of a legendary creature. This bird is known for the remarkable melody it sings in the rain forest once a year. According to the legend, Uirapuru's song is so beautiful that all other birds stop singing to listen to it.
>
> . . .
>
> Consistent with the theme of this year's ICC Biennale, "interaction," Kac's "Uirapuru" explores the interconnectedness of two parallel worlds, physical and virtual, while developing through myriad levels of perception and interaction the poetic of myth and fantasy in the context of complex digital systems.
>
> . . .

"Uirapuru" can be experienced online (until November 28) according
to the following timetable:
Tokyo: between 10 am and 6 pm; Friday until 9 pm.
New York: between 8 am and 4 am; Friday until 6 am.
Rio de Janeiro: between 11 pm and 7 am; Friday until 9 am.
Paris: between 3 am and 11 am; Friday until 1 pm.

Programmed to perform according to signals received from the viewer/lis-
tener/operator, "Uirapuru" would elicit responses based on multiple cultural codes as well as on each viewer's relations to contemporary
technologies. Every interactive performance made the viewer both the
creator and the recipient of a (wordless) "text" that was as unique as it was
irretrievable, while "Uirapuru" remained the work of one artist. The virtual space of the Internet has become the "play area" (Gomringer's "spielraum") in which readers/performers find ever increasing opportunities to
become creative participants in the process of making texts.

Notes

1. The anthologies by year are: Max Bense and Elisabeth Walther, eds.,
   Konkrete poesie international. rot (Stuttgart, 1965) no. 21; *Poesia concreta in-
   ternacional,* catalogue of exhibition at Galeria Universitária Aristos,
   March-May 1966 (Mexico City: Universidad Nacional Autónoma de
   México, 1966); Stephen Bann, ed., Concrete poetry issue, *Beloit Poetry
   Journal* 17,1 (1967); Stephen Bann, ed., *Concrete Poetry:An International An-
   thology* (London: London Magazine Editions, 1967); Eugene Wildman,
   ed., "Anthology of Concretism," *Chicago Review* 19,4 (1966); Emmett
   Williams, ed., *An Anthology of Concrete Poetry* (New York: Something Else
   Press, 1967); Mary Ellen Solt, ed., "A World Look at Concrete Poetry,"
   Artes Hispánicas/Hispanic Arts 1, 3–4 (1968); Adriano Spatola, ed., "An-
   tologia della poesia concreta," *Il peso del concreto,* ed. Ezio Gribaudo
   (Torino, 1968); Carlo Belloli and Ernesto L. Francalanci, eds., *Poesia concreta: indirizzi concreti, visuali e fonetici,* catalogue of exhibition organized by
   Dietrich Mahlow and Arrigo Lora-Totino, Ca' Giustinian, Sala delle
   Colonne, 25 September–10 October 1969 (Venezia: Stamperia di Venezia,
   1969); Max Bense and Elisabeth Walther, eds., *Konkrete poesie international
   2. rot* (Stuttgart, 1970) no. 41; Liesbeth Crommelin, ed., *klankteksten / ?
   konkrete poëzie / visuele teksten* (Amsterdam: Stedelijk Museum, 1970). The
   earliest international anthology using "Concrete Poetry" as the collective
   title was Eugen Gomringer's "Kleine Anthologie konkreter Poesie," *Spi-
   rale* 8 (1960): 37–44. In addition to the three exhibition catalogues above,
   Crommelin's list of "Exhibitions and Manifestations" (228–29) lists catalogues for international Concrete poetry exhibitions in Stuttgart (1959),
   Freiburg (1963 and 1968), Arlington (1968), Rouen (1968), Venice (Bien-

nale, 1969), and Bloomington (Indiana University, 1970). Numerous other exhibitions displaying similar materials used different labels. The international anthologies were followed by larger or smaller collections of *British Concrete Poetry,* ed. John Sharkey (London: Lorrimer, 1971); *Konkrete Poesie: Deutschsprachige Autoren,* ed. Eugen Gomringer (Stuttgart: Philipp Reclam, 1972); and *Poesia Concreta em Portugal,* ed. José Alberto Marques and E. M. de Melo e Castro (Lisboa: Assírio & Alvim, 1973). The Brazilian "Noigandres" group had collectively published its own work as early as 1962 (Augusto de Campos, Décio Pignatari, Haroldo de Campos, José Lino Grünewald, and Ronaldo Azeredo, *Antologia: do verso à poesia concreta, 1949–1962* (São Paulo: Massao Ohno, 1962). Hansjörg Mayer had published *Concrete Poetry: Britain, Canada, United States* in 1966 (Stuttgart, 1966); a Polish anthology of *poezja konkretna: wybór tekstów polskich oraz dokumentacja z lat 1967–1977,* ed. Stanislaw Drózdz, appeared in 1978 (Wroclaw: Socjalistyczny zwiazek studentów polskich akademicki osrodek teatralny kalambur, 1978).

2. A much-quoted event first documented in Augusto de Campos, Décio Pignatari, and Haroldo de Campos, *Teoria da Poesia Concreta: Textos Críticos e Manifestos 1950–1960* (São Paulo: Edições Invenção, 1965), 194.

3. The organizer of the Stedelijk Museum exhibition considered it "obvious" that the movement "has by now become a closed historical chapter" and referred to statements "by various well-known poets in this field" published in 1970 in Nicholas Zurbrugg's magazine *Stereo Headphones* concerning "the 'death' of concrete poetry" (Crommelin, *klankteksten / ? konkrete poëzie / visuele teksten,* n.p.).

4. This is my most recent definition of an intermedia text (in distinguishing it from a multimedia or a mixed-media text), in response to the criticism of my earlier formulation voiced by Jürgen E. Müller, "Intermedialität als poetologisches und medientheoretisches Konzept," *Intermedialität: Theorie und Praxis eines interdisziplinären Forschungsgebiets,* ed. Jörg Helbig (Berlin: Erich Schmidt Verlag, 1998), 38, n7; cf. Claus Clüver, "Interartiella studier: en inledning" ("Interarts Studies: An Introduction"), trans. Stefan Sandelin, in *I musernas tjänst: Studier i konstarternas interrelationer,* ed. Ulla-Britta Lagerroth, Hans Lund, Peter Luthersson, and Anders Mortensen, 17–47 (Stockholm/Stehag: Brutus Östlings Bokförlag Symposion, 1993), 47, n9; original English version unpublished.

5. See for example Craig Saper's exploration of Augusto de Campos' "codigo" in "Under Cancellation: The Future Tone of Visual Poetry," *Experimental—Visual—Concrete: Avant-Garde Poetry Since the 1960s,* ed. K. David Jackson, Eric Vos, and Johanna Drucker (Amsterdam: Rodopi, 1996), 309–16.

6. On this question, see Claus Clüver, "Traduzindo Poesia Visual," *Cânones & Contextos* (Rio de Janeiro: ABRALIC, 1997), 311–27.

7. According to Kathleen McCullough's *Concrete Poetry: An Annotated International Bibliography, with an Index of Poets and Poems* (Troy, NY: Whitston

Publishing, 1989), Niikuni's *"kawa/su"* is among the poems reprinted "in at least twenty-five anthologies and critical works" (xiii).

8. The version, with the title "monotony of void space," was produced for the Williams anthology and reprinted in the Stedelijk Museum catalogue, 81. Haroldo de Campos published a Portuguese version of his translation in his essay "Poesia Concreta no Japão" (*Estado de São Paulo,* Suplemento Literário, 10 May 1958), with a detailed analysis of part 2 in order to show how the principles of Concrete poetry work in a text composed of ideograms; see excerpts in English translation in John Solt, *Shredding the Tapestry of Meaning: The Poetry and Poetics of Kitasono Katue (1902–1978)* (Cambridge: Harvard University Press, 1999), 263–64. See also Haroldo de Campos, "Visualidade e Concisão na Poesia Japonesa" (1964), in *A Arte no Horizonte do Provável e outros ensaios* (São Paulo: Perspectiva, 1969), 63–75. Solt's own translation of "Monotonous Space," admittedly less "flowing," begins: "white square / within it / white square / within it / yellow square . . ."

9. Haroldo de Campos in a note quoted in Williams' anthology, n.p. John Solt states that Katue "was no doubt influenced by painter Josef Albers' series" (359, n13). These are the lines from Kandinsky's "Sehen": " . . . Weißer Sprung nach weißem Sprung. / Und nach diesem weißen Sprung wieder ein weißer Sprung. / Und in diesem weißen Sprung ein weißer Sprung. In jedem weißen Sprung ein weißer Sprung. . . ." ("White leap after white leap. / And after this white leap another white leap. / And in this white leap a white leap. In every white leap a white leap." From Wassily Kandinsky, *Sounds,* trans. Elizabeth R. Napier (New Haven:Yale University Press, 1981). *Sprung* means "leap" or "jump" as well as "crack" (as in a bowl or a wall): the context suggests both meanings.

10. Solt, *Shredding the Tapestry of Meaning,* 359, n12; also 253–59.

11. Ibid., 266.

12. John Solt says nothing about the sound of *"tanchō na kūkan"* but demonstrates the importance of sound in the poems of Katue's *Kuroi hi (feu noir,* 1951); his observations (*Shredding the Tapestry of Meaning,* 238–39) would corroborate my impression about this four-part poem. Solt also reports that "even though Katue had pioneered the modern poetry reading in Japan, debuting on stage in 1935 and reappearing in 1936, he never joined the trend of public poetry readings after the war. He adamantly disapproved of other people's versions of his poems, whether put to music or sung" (238).

13. This has not kept composers from setting them to music. For the case of Brazil, see Claus Clüver, "Concrete Poetry Into Music: Oliveira's Intersemiotic Transposition," *The Comparatist* 6 (1982): 3–15.

14. According to the "Sinopse do movimento da poesia concreta" published at the end of *Teoria da Poesia Concreta: Textos Críticos e Manifestos 1950–1960,* by Augusto de Campos, Décio Pignatari, and Haroldo de Campos, 3rd ed. (São Paulo: Brasiliense, 1987), the first "oralization" of the "Poetamenos" poems (published in 1955) took place in 1954 (193). A 1973 stereo record-

ing of the final poem, "dias dias dias," by Caetano Veloso, was included in Augusto de Campos, *Poesia 1949–1979* (São Paulo: Livraria Duas Cidades, 1979). See also Claus Clüver, "*Klangfarbenmelodie* in Polychromatic Poems: A. von Webern and A. de Campos," *Comparative Literature Studies* 18 (1981): 386–98.

15. See Dick Higgins, "Sound Poetry," *The New Princeton Encyclopedia of Poetry and Poetics*, ed. Alex Preminger and T. V. F. Brogan (Princeton: Princeton University Press, 1993), 1182.

16. Besides all kinds of noises, Jandl's poem introduces commands such as "Exercise!" and "Never fail!" into one of the "quietest" of all German poems: "O'er all the treetops is quiet now . . ." (Longfellow's translation).

17. In print, as in the case of the Goethe transformation, the original is published along with the new version. In a 1984 recording, a reading of the English text by Lauren Newton is followed by Jandl's rendition of his "oberflächenübersetzung," and then both readers recite the two texts simultaneously, accompanied by music composed by Newton (Ernst Jandl, Lauren Newton, Woody Schabata, and Mathias Rüegg, *bist eulen?*, record produced by Harald Quendler and Mathias Rüegg, Extraplatte EX 316 141 (Wien: 1984).

18. Greenham, record notes.

19. Ibid.

20. On the complex question of the status of the subjectivity in experimental poetry, see Friedrich W. Block, *Beobachtung des 'Ich': Zum Zusammenhang von Subjektivität und Medien am Beispiel experimenteller Poesie* (Bielefeld: Aisthesis Verlag, 1999).

21. Haroldo de Campos, *A Operação do Texto* (São Paulo: Perspectiva, 1976); Claus Bremer, "ein text passiert," in *Anlaesse: kommentierte Poesie 1949–1969* (Neuwied und Berlin: Luchterhand, 1970), 29; and Aram Saroyan, "crickets," from "5 Poems/1964–65/New York," *Pages* (New York: Random House, 1969), n.p. See also Bremer's comment: "The reader's activity makes the text pass" ["Die Aktivitaet des Lesers laesst den Text passieren"], 28.

22. According to Umberto Eco, who in 1959 wrote an essay on "Zen e Occidente," by 1968 Western interest in Zen had left noteworthy signs in artistic production only in the United States. A Portuguese translation of his essay is contained in the second Brazilian edition of his *Obra Aberta: Forma e indeterminação nas poéticas contemporâneas*, trans. Giovanni Cutolo (São Paulo: Perspectiva, 1971), 203–25; he singles out the importance of Zen for John Cage, whose work was to provide a major inspiration for the Brazilian Noigandres poets.

23. "Die konstellation wird vom dichter gesetzt. Er bestimmt den spielraum, das kräftefeld, und deutet seine möglichkeiten an, der leser, der neue leser, nimmt den spielsinn auf: dem wissen um die möglichkeiten des spiels kommt heute die gleiche bedeutung zu wie ehedem der kenntnis klassischer dichtersatzungen"; Eugen Gomringer, "vom vers zur konstellation:

zweck und form einer neuen dichtung" (1954–55), reprint in *Theorie der konkreten Poesie: Texte und Manifeste 1954–1997* (Wien: Edition Splitter, 1997), 12–18, quote 16.

24. Originally published as "L'opera in movimento e la coscienza dell'epoca" in *Incontri Musicali* 3 (1959), the essay formed the first chapter in Eco's *Opera Aperta—Forma e indeterminazione nelle poetiche contemporanee* (Milan: Bompiani, 1962). For inclusion in *The Role of the Reader: Explorations in the Semiotics of Texts* (Bloomington: Indiana University Press, 1979), Eco revised Bruce Merry's translation of "The Poetics of the Open Work" published in 1974. An English translation of *Opera Aperta* appeared in 1989 as *The Open Work*, trans. Anna Cancogni (Cambridge: Harvard University Press, 1989).

25. Quoted in Eco, *Role of the Reader,* 48.

26. Eco, *Role of the Reader,* 62.

27. "*Every* work of art, even though it is produced by following an explicit or implicit poetics of necessity [a characteristic of closed works], is effectively open to a virtually unlimited range of possible readings, each of which causes the work to acquire new vitality in terms of one particular taste, or perspective, or personal *performance*" (Eco, *Role of the Reader,* 63). Eco makes a distinction between his use of "open" and "closed" (which he may have derived from Pierre Boulez, see note 29 below) and the application of these terms to contrast Baroque and Renaissance art (by Heinrich Wölfflin, later transferred to literature by Oskar Walzel, whom he does not mention). The quotation is from 56.

28. Haroldo de Campos, "A Arte no Horizonte do Provável," *A Arte no Horizonte do Provável e outros ensaios* (São Paulo: Perspectiva, 1969), 15–32. Originally published in 1963.

29. Haroldo de Campos, "A Obra de Arte Aberta," *Teoria da Poesia Concreta: Textos Críticos e Manifestos 1950–1960,* eds. Augusto de Campos, Décio Pignatari, and Haroldo de Campos, 3rd. ed. (São Paulo: Brasiliense, 1987), 36–39. First published in *Diário de São Paulo* July 3, 1955. "Pierre Boulez, in a conversation with Décio Pignatari, declared his lack of interest in the "perfect," "classical" work of art "of the diamond type" and explained his concept of the *open work of art,* like a "modern baroque" (*Teoria* 39, my trans.).

30. *The I Ching or Book of Changes,* the Richard Wilhelm translation rendered into English by Cary F. Baynes (Princeton: Princeton University Press, 1967), 189–92.

31. Jean M. Borgatti and Richard Brilliant, *Likeness and Beyond: Portraits from Africa and the World* (New York: The Center for African Art, 1990).

32. Dietrich Mahlow, ed., *Schrift und Bild / Schrift en beeld / Art and Writing / L'art et l'écriture,* catalogue for exhibition at Stedelijk Museum, Amsterdam, May 3-June 10, 1963, and at Staatliche Kunsthalle Baden-Baden, June 15-August 4, 1963 (Frankfurt: Typos Verlag, 1963); *Between Poetry and Painting,* exhibition catalogue, Institute of Contemporary Arts, London, October

22-November 27, 1965 (London: W. Kempner, 1965); Felix Andreas Baumann, org., *Text Buchstabe Bild*, exhibition catalogue, Helmshaus Zürich, July 11-August 23, 1970 (Zürich: Zürcher Kunstgesellschaft, 1970).

33. Bernard Blistène and Véronique Legrand, orgs., *Poésure et Peintrie: «d'un art, l'autre»*, exhibition catalogue, Centre de la Vieille Charité, Marseille, February 12-May 23, 1993 (Marseille: Réunion des Musées Nationaux, Musées de Marseille, 1998).

34. Suzanne Pagé, org., *Écouter par les yeux: objets et environnements sonores*, exhibition catalogue (Paris: ARC, Musée d'Art Moderne de la Ville de Paris, 1980).

35. Dietrich Mahlow, ed., *Auf ein Wort! Aspekte visueller Poesie und visueller Musik*, exhibition catalogue, Gutenberg-Museum, Mainz, 1987 (Mainz: Edition Braus, 1987).

36. Thomas Kellein, *«Fröhliche Wissenschaft»: Das Archiv Sohm*, exhibition catalogue, Staatsgalerie Stuttgart, November 22, 1986-January 11, 1987 (Stuttgart: Staatsgalerie, 1986).

37. Johanna Drucker, "Experimental, Visual, and Concrete Poetry: A Note on Historical Context and Basic Concepts," in *Experimental—Visual—Concrete: Avant-Garde Poetry Since the 1960s*, ed. Jackson, Vos, and Drucker, 39–61.

38. The variety of "visual poetics/poetries" developed by 1977 is illustrated in the catalogue for the international exhibition assembled at the Museu de Arte Contemporânea da Universidade de São Paulo by Walter Zanini and Julio Plaza. A very recent compilation, in Russian, is Dmitry Bulatov's 591-page *A Point of View: Visual Poetry: The 90s, An Anthology* (Königsberg: Simplisii, 1998).

39. Klaus-Peter Dencker, ed., *Visuelle Poesie aus Japan: Eine Ausstellung der Kulturbehörde der Freien und Hansestadt Hamburg*, catalogue (Hamburg: Kulturbehörde, 1997).

40. Hiroo Kamimura, "En hommage à Gertrude Stein," color reproduction in *Visuelle Poesie aus Japan*, ed. Dencker, 61. Kamimura is one of the most important mediators between Japanese and Western, especially German, visual poets.

41. Chiharu Nakagawa, *Alibi Ware (Poetry)*, color reproduction in *Visuelle Poesie aus Japan*, ed. Dencker, 83.

42. See the illustrations from Kriwet's *Textroom* in *Imaged Words & Worded Images*, ed. Richard Kostelanetz (New York: Outerbridge & Dienstfrey/Dutton, 1970), 14–18. The book also contains (50–57) views of John Furnival's much reproduced *Tours de Babel Changées en Ponts* (1964), described by Kostelanetz as "perhaps the most spectacular example in the worded-image strain . . . , in which legible words in several languages are piled into shapes, particularly towers, over six panels—each in the original version over six feet high, with the work as a whole running over twelve feet across" ("Introduction," n.p.).

43. In Emmett Williams' *Sweethearts* (New York: Something Else Press, 1967), all pages of the erotic poem cycle contain texts formed by letters contained

in a square formed by eleven repetitions of the eleven-letter word "sweet-hearts"; some of the pages also form "kinetic metaphors" that may be created "by flipping the pages fast enough to obtain a primitive cinematic effect" (E. W.). In his *A Valentine for Noel: Four Variations on a Scheme* (Barton, VT: Something Else Press, 1973), the same flipping technique must be used to constitute three of the texts.

44. E. M. de Melo e Castro, *Poética dos Meios e Arte High Tech* (Lisboa: Vega, 1988); reprinted in part as "Uma Rede Intersemiótica" in his *O Fim Visual do Século XX*, ed. Nádia Batella Gotlib (São Paulo: EDUSP, 1993), 215–43.

45. Eduardo Kac, ed., *New Media Poetry: Poetic Innovation and New Technologies*, topical issue, *Visible Language* 30,2 (1996); Eric Vos, "New Media Poetry: Theory and Strategies," *New Media Poetry* 214–33.

46. 3.13/14/15/16 (1997), with CD-ROM *Alire* 10 (May 1998).

47. Michael Webster's *Reading Visual Poetry after Futurism: Marinetti, Apollinaire, Schwitters, Cummings* (New York: Peter Lang, 1995) sheds some light on the performance aspects of these poets' work. For the Russian scene, see Gerald Janecek, *Zaum: The Transrational Poetry of Russian Futurism* (San Diego: San Diego State University Press, 1996), which includes indications of performance practices; it is a sequel to Janecek's *The Look of Russian Literature: Avant-Garde Visual Experiments, 1900–1930* (Princeton: Princeton University Press, 1984).

48. Mel Gordon, ed., *Dada Performance* (New York: PAJ Publications, 1987).

49. Cf. Gerhard Rühm's account of the early 1950s, "Vorwort," *Die Wiener Gruppe: Achleitner, Artmann, Bayer, Rühm, Wiener. Texte, Gemeinschaftsarbeiten, Aktionen*, ed. Gerhard Rühm (Reinbek bei Hamburg: Rowohlt, 1967), 7–36.

50. Bob Cobbing and Peter Mayer's *concerning concrete poetry* (London: Writers Forum, 1978), contains "A Chronology of Visual Poetry" (63–70) and "A Chronology of Sound Poetry" (71–74), both based on research by Dom Sylvester Houédard.

51. The festivals have been documented in the following: Fernando Aguiar and Manoel Barbosa, orgs., *PerformArte: I Encontro Nacional de Performance*, exhibition catalogue, April 13–28, 1985, Torres Vedras (Portugal); Fernando Aguiar, org., *1.º Festival Internacional de Poesia Viva*, exhibition catalogue, April and May, 1987, Figueira da Foz (Figueira da Foz, Portugal: Museu Municipal Dr. Santos Rocha, 1987); Fernando Aguiar, org., *Il Encontro Nacional de Intervenção e Performance*, exhibition catalogue, July 9–August 7, 1988, Amadora (Portugal), Galeria Municipal/Recreios Desportivos (Lisboa: Associação Poesia Viva, 1988). See also: Fernando Aguiar and Gabriel Rui Silva, orgs., *Concreta. Experimental. Visual: Poesia Portuguesa 1959–1989*, exhibition catalogue, April 10–17, 1989, Università di Bologna, Palazzo Herculani (Lisboa: Instituto de Lingua e Culture Portuguesa, 1989).

52. "Shimamoto Shozo," *Visuelle Poesie aus Japan*, ed. Dencker, 93–96, with three illustrations.

53. Arnaldo Antunes, *NOME,* with Celia Catunda, Kiko Mistrorigo, and Zaba Moreau (São Paulo: BMG Ariola Discos, 1993). Music videocassette plus book and brochure with English version of verbal texts.

54. Richard Kostelanetz, ed., *Text-Sound Texts* (New York: William Morrow, 1980).

55. Henry M. Sayre, *The Object of Performance: The American Avant-Garde since 1970* (Chicago: University of Chicago Press, 1989); cf. Marjorie Perloff, *The Poetics of Indeterminacy: Rimbaud to Cage* (Princeton: Princeton University Press, 1981).

56. For a richly illustrated introduction, see Yves Abrioux, *Ian Hamilton Finlay: A Visual Primer,* introd. notes by Stephen Bann, rev. and expanded ed. (Cambridge: MIT Press, 1992), esp. 38–70.

57. Kac has written extensively about his work in this field. For a short introduction, see Eduardo Kac, "Holopoetry and Hyperpoetry," in *The Pictured Word,* ed. Martin Heusser, Claus Clüver, Leo Hoek, and Lauren Weingarden (Amsterdam: Rodopi, 1998), 169–79.

Chapter 3

Vulfolaic the Stylite: Orientalism and Performing Holiness in Gregory's *Histories*

Karen A. Winstead

"Syria," Peter Brown observed in one of his many brilliant essays on late antique spirituality, was known in the West as "the great province for ascetic stars."[1] Perhaps the most dazzling of those cynosures was Simeon Stylites (A.D. 390–459), whose reputation for flamboyant self-mortification mounted in step with his physical elevation: he stood atop a series of ever-taller pillars for the last thirty years of his life and died sixty feet above the ground, a cult figure in his own lifetime.[2] Simeon's example spawned generations of imitators. As Hippolyte Delahaye wrote in his 1923 study of the stylite movement:

> Pendant de longs siècles l'héroïque extravagance du grand Syméon exerça une véritable fascination sur l'esprit des ascètes orientaux; et malgré les difficultés matérielles qu'entraîne le séjour dans un ermitage élevé au-dessus de terre, le nombre de stylites qui sont nommés dans l'histoire ecclésiastique est véritablement étonnant. Il y a plus. Les textes ne manquent pas où les stylites sont cités comme formant une catégorie à part; le nom désigne une élite de moines relativement nombreuse et fort considérée.[3]

> [For centuries, the heroic extravagance of the great Symeon captured the imagination of Eastern ascetics; and despite the material difficulties involved in occupying a hermitage raised off the ground, ecclesiastical history identifies a truly astonishing number of stylites. There is more. There is no lack of references to stylites as a category unto themselves; the name designated a monastic elite that was relatively numerous and highly regarded.]

That Saint Simeon had many imitators is hardly surprising, for holiness
was, by definition, imitation. Aspiring saints reenacted the lives of earlier
saints, which were in turn modeled on that of the archetype of holiness,
Christ. Hagiographers, for their part, repeated the same plots *ad infinitum*.
They established sanctity by drawing parallels between the lives of their
subjects and those of famous predecessors, frequently designating their
heroes "a second X" or "a new Y," where X and Y were famous figures
from the Bible or from Christian legend. The sixth-century bishop Gregory of Tours, one of the most prolific hagiographers of the early Middle Ages and one of hagiography's earliest theorists, insisted that it is
more accurate to speak of the life (*vita*) of the saints than of the lives
(*vitae*) of the saints.[4]

Though most scholars once dismissed hagiography as lacking historical
and literary merit, deploring its preference for factitious cliché over "authentic" detail, recent cultural historians have mined the genre for evidence of the beliefs and values of the communities in which saints' legends
circulated.[5] Equally illuminating—though rarer than saints' lives—are the
stories of men and women whose enactments of sacred paradigms go awry,
for their stories reveal dissent within ideologies that might otherwise seem
universal. This essay examines one such story, that of Vulfolaic, the would-
be stylite of Carignan, as it is related by Gregory of Tours. Reenacting the
public, heroic asceticism of the Eastern saint Simeon Stylites, Vulfolaic
attempted to participate in the *vita sanctorum*, only to be told that his chosen mode of *imitatio* could not be practiced in the West; obediently, he re-
cast himself in an approved supporting role—keeper of relics of the saintly
dead—which Gregory himself followed. Vulfolaic's metamorphosis from
an unappreciated "new Simeon Stylites" into a successful "new Gregory of
Tours" is an instructive commentary on how holiness was, and wasn't, supposed to be performed in sixth-century Gaul and offers, as well, an early
historical example of some of the dilemmas of cross-cultural performance,
especially as it moves across the East-West divide.

In Book 8 of his *Histories,* Gregory reports meeting a certain Vulfolaic
(also variously known as Vulfilaic, Wulfilach, and Saint Walfroy) on a chance
visit to the latter's monastery just outside Carignan.[6] Though now an em-
inently respectable deacon, Vulfolaic had once been an itinerant idoloclast
who, on settling down "in the neighbourhood of Trier," was dismayed to
find the "credulous locals" worshipping a statue of Diana (446). Vulfolaic
erected a column on which he stood day and night, drawing crowds to the
spectacle of his asceticism, and his eloquent preaching persuaded the erstwhile devotees of Diana to topple and smash their goddess. Despite this
success, Vulfolaic's career as a stylite was curtailed by "certain bishops," who
protested:

It is not right, what you are trying to do! Such an obscure person as you can never be compared with Simeon the Stylite of Antioch! The climate of the region makes it impossible for you to keep tormenting yourself in this way. Come down off your column, and live with the brethren whom you have gathered around you.[7]

Vulfolaic obeyed, descending from his column and joining communal life. The bishops' objections seem curiously inappropriate, given Simeon's career. The renowned stylite was himself the obscure son of a shepherd until he achieved a self-made celebrity, and Vulfolaic was well on the way to doing the same: "Crowds began to flock to me from the manors in the region" (446). As for the climate, Simeon's own fame rested, in part, on his heroic endurance of the elements,[8] and Vulfolaic had already endured a Gallic winter. The climate really at issue here is the politico-religious climate of sixth-century Gaul: The bishops are upset because Vulfolaic has imported a distinctly Eastern model of holiness.

As Peter Brown has shown, Eastern and Western Christendom had, by Gregory's time, developed radically different paradigms of sanctity. The East teemed with living saints such as Simeon, who were known for ostentatious self-mortification. Though revered and admired, these saints were social outsiders:

> The Christian *koiné* was articulated in the eastern Mediterranean not to place the holy man above human society, but outside it. . . . The East Roman holy man, as we observe him in both the fourth and fifth centuries, and in the age of Gregory . . . [,] preserved his reputation . . . by an exacting ritual of de-solidarization and even of social inversion. He wielded his "idealized" power by adopting stances that were the exact inverse of those connected with the exercise of real power.[9]

The powers accorded to Eastern saints were extensive and varied: they settled lawsuits, mediated disputes, cursed erring potentates, cured the sick, and regulated the weather. In the West, by contrast, saints were "safely dead"; God's power was located in their remains, and the authority accruing to that power was claimed by the guardians of those remains, the bishops.[10] In the East, anyone with enough stamina and charisma could attain the status of holy man; in the West, saints and bishops alike were for the most part well pedigreed. Gregory himself boasts that "apart from five, all the other bishops who held their appointment in the see of Tours were blood-relations of my family" (321). Western charismatics of the sort venerated in the East were, for Gregory and his cohorts, at best crazies and at worst frauds. An example of this attitude is found in Book 10 of the *Histories:*

> Quite a number of men now came forward in various parts of Gaul and by
> their trickery gathered round themselves foolish women who in their frenzy
> put it about that they were saints. These men acquired great influence over
> the common people. I saw quite a few of them myself. I did my best to
> argue with them and to make them give up their inane pretensions. (586)

The bishops who ordered Vulfolaic off his column were engaged in the
same endeavor. If Vulfolaic fancied himself a new Simeon, the bishops saw
in him a new Diana. Not content with deposing Vulfolaic, they toppled his
column and smashed it with hammers, reenacting Vulfolaic's demolition of
the idol.

Although they venerated canonical Eastern saints, including Simeon,
Western ecclesiastics of Gregory's day looked askance at the entrepreneur-
ial character of Eastern holiness, seeing it as inimical to the establishment
of their authority vis-à-vis secular potentates and the public at large. The
Histories abound with proof that, whether aristocrats or commoners, those
who obey their bishops are good, while those who malign, challenge, or
otherwise mistreat their bishops are bad; those who donate are pious, while
those who resent episcopal wealth are "evil" (350, 380).[11] Gregory hammers these points home in countless exempla. Thus, we learn of Count
Nantinus of Angoulême, who died shrieking that a bishop he had once
wronged in "the most unpardonable way" was torturing him to death.
When he finally "breathed out his unhappy soul," his body turned black as
coal, "incontrovertible proof that he was being punished to avenge the
holy Bishop" (300). Lest we miss the point, Gregory supplies a moral: "All
should stand amazed and be filled with awe at what happened, and more
especially all should take care not to offend their bishops, for the Lord will
avenge his servants if they put their trust in Him" (301).

Vulfolaic's story could be read as yet one more illustration of Gregory's
claim that "our God is ever willing to give glory to His bishops" (265). The
stylite immediately bows to the authority of the bishops, iterating Gregory's favorite theme: "[I]t is considered a sin not to obey bishops, so, of
course, I came down"; "I have never dared to set up again the column
which they broke, for that would be to disobey the commands of the bishops" (447). Completing the bishops' triumph, he adopts a quintessentially
Western mode of holiness, that of proprietor of relics. Vulfolaic has already
built a large church and enshrined there some dust from the tomb of Saint
Martin of Tours. He now throws his energies into promoting his church,
making it "famous for its relics of Saint Martin and other saints" (445). By
the time Gregory meets him, Vulfolaic is reluctant to speak of his own past
achievements but eager to advertise the miracles of his patron: a deaf-mute
boy cured while sleeping in the church, an arsonist struck dead for perjury,

and many more. Vulfolaic figures prominently in these *miracula*. Saint Martin personally informs him, in a vision, of the boy's cure, while in the perjury incident, Vulfolaic occupies center stage:

> As he [the arsonist] made to come in to take the oath, I went to meet him. "Your neighbors assert that, whatever you say, you cannot be absolved from this crime," I said. "Now, God is everywhere, and His power is just as great outside the church as it is inside. If you have some misguided conviction that God and His Saints will not punish you for perjury, look at His holy sanctuary which stands before you. You can swear your oath if you insist; but you will not be allowed to step over the threshold of this church." (448–449)

The sinner's brazen defiance—"He raised his hands to heaven and cried: 'By Almighty God and the miraculous power of His priest Saint Martin, I deny that I was responsible for this fire'"—makes the fulfillment of Vulfolaic's prophecy all the more dramatic:

> As soon as he had sworn his oath, he turned to go, but he appeared to be on fire himself! He fell to the ground and began to shout that he was being burnt up by the saintly Bishop. In his agony he kept shouting: "As God is my witness, I saw a flame come down from heaven! It is all around me and it is burning me up with its acrid smoke!" As he said this, he died. (449)

Vulfolaic's flair for spectacle has not diminished since his days as a stylite, but the nature and significance of the spectacle are completely different. In displaying his rapport with the holy dead, Vulfolaic follows a Western ecclesiastical route to authority.

One could in fact argue that Gregory presents the deacon's story as a fairly straightforward, normative progression toward Western Christianity. Early on, Vulfolaic converts ignorant pagans into nominal Christians, but he does so by quite literally taking Diana's place atop the pedestal. Then the bishops, agents of order and stability, finish what Vulfolaic has begun by transforming the well-intentioned but misguided stylite into a respectable pastor. So perfect is Vulfolaic's conversion that Gregory meets in him a veritable double: a builder of churches, a teller of miracle stories, and a custodian of relics. Indeed, he advertises the same kind of miracles that Gregory records by the dozen in his own *libri miraculorum*. "Famous for its relics of Saint Martin and other saints" (445), Vulfolaic's church is, in the parlance of hagiographers, a new Tours.

However, not all details of Vulfolaic's story fit into this tidy picture. Vulfolaic, even as a respectable deacon, still feels that he was wrongly ordered off his column: "There came to me certain bishops whose plain duty it was to exhort me to press wisely on with the task which I had begun. Instead they [abjured him from it]" (447).[12] In fact, he goes so far as to attribute the bishops' interference to infernal conniving: "And because the envious one always tries to harm those seeking God, certain bishops came. . . ."[13] The bishops do not convince Vulfolaic that he is in error, as Vulfolaic had done with Diana's devotees, but rather operate through naked authority— ordering him off his column, in the passage quoted earlier—combined with subterfuge: "One day a certain bishop persuaded me to go off to a manor which was some distance away. Then he sent workmen with wedges, hammers and axes, and they dashed to pieces the column on which I used to stand." Vulfolaic's conclusion, "I wept bitterly, but I have never dared to set up again the column which they broke, for that would be to disobey the commands of the bishops," could be read as ironic resignation rather than true submission.

As I suggested earlier, the medieval reader of hagiography, accustomed to analyzing the repetition and variation of familiar themes within and among saints' lives, might see the bishops as taking up where Vulfolaic left off in deposing idols, with the ironic twist that Vulfolaic himself has become the idol. But the same events invite a competing interpretation: the bishops' demolition of Vulfolaic's pillar could be read as a parodic reenactment of Vulfolaic's destruction of Diana's statue by misguided men who don't understand that one is supposed to smash only pagan monuments, not those erected to God. Gregory would surely not have endorsed this interpretation, but its availability hints that he was not entirely comfortable with the bishops' victory.

Gregory himself does not come off so well in the Vulfolaic episode: "At first he was unwilling to tell me his story, for he was very sincere in his desire to avoid notoriety. I *adjured him with terrible oaths,* begging him not to keep back anything of what I was asking him, and *promising never to reveal what he told me to a living soul*" (my emphasis, 445).[14] Indeed, one might detect just a hint of professional jealousy on Gregory's part toward a man whom he somewhat condescendingly refers to as "a Longobard [Lombard] by birth" (445). Vulfolaic, after all, is a nobody, who began his service to Saint Martin not knowing whether Martin "was a martyr or just a famous churchman" or even "what good he had done in this world" (445). What's more, he confesses to Gregory, he had no idea "which place had been honoured by receiving his sacred body for burial," i.e., he was unfamiliar with the church at Tours, whose fame Gregory had invested so much of his career in promoting.[15] Yet from a small box of dust gathered from Martin's tomb, Vulfolaic succeeds in creating not just a new Tours but in some ways a better one. While Vulfolaic appears to enjoy Saint Martin's unfailing protection, Gregory's *Histories* recount numerous atrocities committed at his own church of Saint Martin. Rogues steal the saint's possessions, cheat his bishops, perjure themselves on his relics, and desecrate his sanctuary. Retribution for these impieties is often long in coming—if, indeed, it comes at all. In a showdown similar to that of Vulfolaic with the arsonist, the villain Merovech arrives at the church of Saint Martin as Gregory is celebrating Mass and demands Holy Communion. Gregory at first refuses but backs down when Merovech threatens to kill members of the congregation.[16] Where, readers might well wonder, was Saint Martin? Martin was also "largely absent," as Ian Wood has observed, in the great crisis of Gregory's career, his trial for calumniating Queen Fredegund, a charge laid against him by one who "had sworn loyalty to me three or more times on the tomb of Saint Martin" (317). Gregory's enemies clearly were "less than impressed by the bishop's special relationship with his supernatural patron," as Wood puts it.[17] Adding insult to injury, Gregory finds, on returning to Tours after his narrow escape, that his perfidious archdeacon Riculf has usurped Saint Martin's Church and now scorns Gregory's authority, threatening, indeed, to kill him. Although Gregory manages to reassert his position, Riculf not only escapes punishment but finds a warm welcome in another diocese.

The reconstructed Vulfolaic might well be the envy of many bishops besides Gregory. There is more of Gregory vs. Merovech than of Vulfolaic vs. the arsonist in the performance of Bishop Cautinus of Clermont, who in Book 9 confronts the notorious rogue and suspected matricide Count Eulalius as he is about to take communion in the church of Saint Julian— on Julian's feast day, no less.[18] At the altar, Cautinus cautions Eulalius:

> It is common talk among the people that you killed your own mother. I do not know whether or not you really committed this crime. I therefore leave it to God and to the blessed martyr Julian to judge this matter. If you really are innocent, as you maintain, draw near, take your portion of the consecrated bread and place it in your mouth. God will be looking into the deepest confines of your heart. (554)

What followed surely diminished both Cautinus and Saint Julian in the eyes of the congregation: "Eulalius took the consecrated bread, communicated and went his way"—for although Cautinus professes an open mind, Gregory furnishes a litany of Eulalius' subsequent outrages that strongly implies that he has with impunity conjoined sacrilege to murder.

Gregory certainly may have intended the Vulfolaic episode to furnish a simple exemplum exhorting strict obedience to bishops. This, indeed, was Edward James's reading: "The story . . . was designed specifically to show how ascetics must obey their bishops."[19] As I have argued, however, the successes of the quixotic Vulfolaic draw attention to the all-too-frequent failures of the ideals and rituals of holiness that Gregory cherishes. Why, then, did Gregory include it at all? Or, perhaps more interesting, what does his inclusion and development of the episode reveal of his implicit assumptions about holiness?

We might begin by considering why Vulfolaic succeeds where Gregory and his episcopal colleagues so often fail. One obvious answer is that the incidents Vulfolaic relates are generically marked as miracle stories.[20] Gregory introduces Vulfolaic's stories by reporting that he had asked the deacon to tell him some of the miracles Saint Martin performed at his church. Of course, the deacon reports only his successes, just as Gregory reports only successes in his *libri miraculorum*. Yet Vulfolaic's accomplishments clearly indicate a special rapport with God. Perhaps Gregory related Vulfolaic's "stylite phase" followed by his "deacon phase" simply to contrast wrongheaded and acceptable ways of performing Christianity; or perhaps Gregory presents both phases because, in his mind, Vulfolaic's later success as the proprietor of Saint Martin's Church is inextricably linked with his earlier stylitism.

To a certain extent, Gregory shared Vulfolaic's fascination with Simeon Stylites, devoting to him a chapter in one of his books of miracles, the *Glory of the Confessors*.[21] That chapter, as Raymond Van Dam points out, "seems quite out of place sandwiched between chapters about a sixth-century abbot of Tours and a third-century bishop of Limoges."[22] Indeed, Simeon is the only Eastern saint whose miracles Gregory treats, and unlike most of Gregory's subjects, he has no apparent connection with Gaul.

Gregory's treatment of Simeon nonetheless reveals some uneasiness, for he does all he can to erase the exotic, Eastern flavor of Simeon's sanctity. He tucks the saint's stylitism into subordinate clauses: "The confessor Simeon, who is said to have stood on a pillar in the region of Antioch, often offered cures to the inhabitants."[23] Likewise: "But in his zeal for holiness after he placed himself on a higher pillar, he did not allow himself to be seen not only by an unrelated woman but even by his own mother."[24] Gregory's sentence-level subordination of stylitism reflects the overall structure of the brief chapter, which dwells not on stylitism but on an approved virtue of Western holy men: avoidance of women. Simeon the Stylite thus becomes, in Gregory's hands, Simeon the Celibate, who stood on pillars.

Gregory grants Simeon's holiness, but feels compelled to naturalize it; Vulfolaic feels no such compulsion. He performs the *imitatio* and enacts the *vita sanctorum* that Gregory merely theorizes about. For Gregory, living the *vita* is problematic. His vast hagiographic oeuvre contains only one short collection of saints' lives, and even in those lives, Gregory seems more comfortable with posthumous miracles than with the deeds of living saints.

His decision to relay Vulfolaic's past through the deacon's first-person narration may reflect the same discomfort with even narrating holy lives. Eastern and Western manifestations of the *vita sanctorum* are, however, equally available to Vulfolaic, and are pursued with a naive enthusiasm. Gregory seems alive to the possibility that this simplicity of spirit, which he cannot bring himself to envy, might nonetheless have earned Vulfolaic a privileged position before God.

While implying that Vulfolaic may be a holier man than the bishops who ordered him off his pillar, and, for that matter, a holier man than he himself, Gregory scarcely implies that he and his colleagues should emulate the erstwhile stylite. Although he shows that the bishops were less than honest with Vulfolaic, everything we know about the bishop of Tours indicates that, had he been on the spot, he would have taken their part. For Gregory, doing God's work in Merovingian Gaul entails increasing God's wealth and using that wealth to do His work, upholding the sanctity of God's churches and maintaining the jurisdiction of His clergy, and protecting the faithful from heretics and even, if need be, from well-intentioned but misguided servants of God. Execution of these duties requires pragmatism, political savvy, and willingness to compromise—touches, in short, of worldly contamination—and it requires that the power of the sacred be concentrated in the hands of those of God's agents who can wield it best. The *Histories,* with the object of recording the "muddled and confused" (103) course of human events, allowed Gregory both to acknowledge the genuine holiness of a Vulfolaic and to insist that, in a fallen world, God's business cannot be accomplished through holiness alone.

It may be true that Gregory, in his portrayal of Vulfolaic, is expressing an implicit view of Eastern ascetic practices as crude and perhaps a touch barbaric. Yet we must not see Gregory as simply deprecating such Eastern models. After all, he does not condemn Vulfolaic's original course; indeed, he allows Vulfolaic to make a persuasive case for it. Stylitism itself is not necessarily the major issue. Throughout his oeuvre, Gregory displays a clear preference for a bureaucratic, organizational Christianity over an heroic model based on imitation. At a time when the Church is threatened not by paganism but by renegade Christians who covet its property and prerogatives, God's interests are best served not by imitating the saints—thus competing with the clergy and complicating their jobs—but by tending the saints' shrines, reporting their miracles, and deploying the power of the holy dead against living villains. That Vulfolaic must be adjured from a flagrantly Eastern form of *imitatio* should not, in other words, obscure the possibility that what is out of place in Gregory's Gaul is not stylitism but *imitatio* itself.

Notes

1. "The Rise and Function of the Holy Man in Late Antiquity," *Society and the Holy in Late Antiquity* (Berkeley and Los Angeles: University of California Press, 1982), 109.
2. "The Life of St. Simeon Stylites: A Translation of the Syriac Text in Bedjan's *Acta Martyrum et Sanctorum*, Vol. IV," *Journal of the American Oriental Society* 35 (1915): 103–198. Robert Doran also provides translations of the Greek versions by Theodoret and Antonius in *The Lives of Simeon Stylites* (Kalamazoo: Cistercian Publications, 1992).
3. Hippolyte Delehaye, *Les Saints Stylites* (Brussels: Société des Bollandistes, 1923), cxvii.
4. Gregory argues this point on grammatical grounds but, as Edward James maintains, certainly has a typological intent. See *The Life of the Fathers,* trans. Edward James, 2nd edition (Liverpool: Liverpool University Press, 1991), xiv and 2.
5. For an analysis and overview of recent approaches, see Kathleen Ashley and Pamela Sheingorn, "Introduction: Reading Hagiography," in their *Writing Faith: Text, Sign, and History in the Miracles of Sainte Foy* (Chicago: University of Chicago Press, 1999), 1–21.
6. References are to and quotations are from the following texts: Gregory of Tours, *The History of the Franks,* trans. Lewis Thorpe (New York: Penguin, 1974); and *Gregorii Episcopi Turonensis Historiarum Libri X,* ed. Bruno Krusch, Monumenta Germaniae Historica: Scriptores Rerum Merovingicarum, Vol. 1 (Hanover, 1885). The Vulfolaic chapters are found, respectively, on 445–49 and 380–84.
7. "Non est aequa haec via, quam sequeris, nec tu ignobilis Symeoni Anthiochino, qui colomnae insedit, poteris conparare. Sed nec cruciatum hoc te sustenere patitur loci posito. Discende potius et cum fratribus, quos adgregasti tecum, inhabita" (*Gregorii Episcopi Turonensis Historiarum Libri X,* 382–83).
8. "The Life of St. Simeon Stylites," 155.
9. Brown, "Eastern and Western Cristendom in Late Antiquity: A Parting of the Ways," *Society and the Holy,* 181–82.
10. Ibid., 187.
11. For a discussion of Gregory's use of cults in the service of episcopal authority, see Raymond Van Dam, "Gregory of Tours and His Patron Saints," in his *Saints and Their Miracles in Late Antique Gaul* (Princeton: Princeton University Press, 1993), 50–81.
12. The Latin reads, "advenientibus episcopis, qui me magis ad hoc cohortare debuerant, ut coeptum opus sagaciter explicare deberem" (382).
13. The Latin, which Thorpe does not translate, reads, "Et quia semper ipse invidus Deum quaerentibus nocere conatur . . ." (382).
14. "Quem ego terribilibus sacramentis coniurans, pollicitus primo, ut nulli quae referebat expanderem, rogare coepi, ut nihil mihi de his quae interrogabam occoleret" (380).

15. For more on Gregory's role in the development of Martin's cult, see Sharon Farmer, *Communities of Saint Martin: Legend and Ritual in Medieval Tours* (Ithaca: Cornell University Press, 1991), 26–29; and Van Dam, "Different Saints, Different Cults," in *Saints and Their Miracles in Late Antique Gaul*, 13–28.

16. This incident is related in *Histories* 5.14.

17. *The Merovingian Kingdoms, 450–751* (London: Longman, 1994), 87.

18. See my discussion of this episode as one of several "abortive miracle stories" in "The Transformation of the Miracle Story in the *Libri Historiarum* of Gregory of Tours," *Medium Aevum* 59 (1990): 1–15.

19. "Introduction," *Life of the Fathers*, xv.

20. See Winstead, "Transformation," 9–11.

21. Gregory of Tours, *Glory of the Confessors*, trans. Raymond Van Dam (Liverpool: Liverpool University Press, 1988), 41; *Gregorii Turonensis Opera: Miracula et Opera Minora*, ed. W. Arndt and B. Krusch (Hanover, 1885), 764.

22. "Introduction," *Glory of the Confessors*, 8.

23. "Symeon vero confessor, qui in columna pagi Anthiocensis dicitur stetisse, frequenter incolis tribuit sanitates."

24. "[P]ostquam vero colomnae editiori se sanctitate fervens invexit, non modo extraneae mulieri, verum etiam nec propriae matri se videndum permisit."

Part II

Occident Meets Orient:
Nation, State, and Local Tradition

Chapter 4

"Strange and Exotic":
Representing the Other in
Medieval and Renaissance Performance

Kathleen M. Ashley

The discourse of the other has become a recognizable mark of contemporary cultural theory, as C. Clifford Flanigan pointed out over
a decade ago: "One of the most important aspects of late twentieth-century social theory and cultural history is the discovery of heterol-
ogy, or the recognition of different and anti-structural elements in various
forms of social and cultural production."[1] Drawing variously on Lacanian
theory, Derridean deconstruction, Marxist poststructuralism, and anthropological models, the discourse of the other has been put to widespread
use in cultural and postcolonial studies. Recent analyses of medieval and
early modern performance, too, are marked by their recognition of otherness in the form of the Jew, the Moor, the "carnivalesque," and so on. In
this essay I will focus on the highly charged scenes of the strange or exotic
played in medieval and early modern performance through characters and
actions distanced in time and space, teasing out the effects of sensation and
spectacle such scenes imply. I want to linger more than is usually permis-
sible on the ways in which these performance moments offered an alternative, pleasurable experience of otherness to their audiences, and thus to
explore the formidable powers of surprise and wonder.

That the exotic is a significant category of representation in early performance has not been widely recognized, perhaps because the dominant
paradigms ruling interpretation of the Middle Ages tend to deny the
power of otherness. The Western discourse of the other from the medieval
to the modern period is usually interpreted as a rhetoric of control and

colonial domination.[2] The representative other is seen as part of a mechanism of "negative self-definition," as Peter Mason puts it.[3] Paul Vanden-
broeck's study of European wildmen and -women, fools, peasants, and
beggars shows that all are types of what the patron or consumer of the visual or textual image is not.[4] Their wildness or deviance defines by inversion the dominant class' normative civilization.

Taking a slightly different tack that nevertheless also implies a negative
view of the other, anthropologist Johannes Fabian describes the medieval
Judeo-Christian vision of Time as a hegemonic view of temporal relations:
"In the medieval paradigm, the Time of Salvation was conceived as inclu-
sive or incorporative," he argues. "The Others, pagans and infidels (rather
than savages and primitives), were viewed as candidates for salvation. Even
the conquista, certainly a form of spatial expansion, needed to be propped
up by an ideology of conversion. One of its persistent myths, the search for
Prester John, suggests that the explorers were expected to round up, so to
speak, the pagan world between the center of Christianity and its lost pe-
riphery in order to bring it back into the confines of the flock guarded by
the Divine Shepherd."[5]

Another typical view of the medieval management of the other is made
by Wlad Godzich, who argues that:

> Western thought has always thematized the other as a threat to be reduced,
> as a potential same-to-be, a not-yet-same. The paradigmatic conception here
> is that of the quest in romances of chivalry in which the adventurous knight
> leaves Arthur's court—the realm of the known—to encounter some form of
> otherness, a domain in which the courtly values of the Arthurian world do
> not prevail. The quest is brought to an end when this alien domain is
> brought within the hegemonic sway of the Arthurian world: the other has
> been reduced to (more of) the same. The quest has shown that the other is
> amenable to being reduced to the status of the same. And, in those few in-
> stances where the errant knight—Lohengrin, for example—does find a
> form of otherness that he prefers to the realm of the same from which he
> came, this otherness is interpreted—by contemporary critics as much as by
> medieval writers—as the realm of the dead, for it is ideologically incon-
> ceivable that there should exist an otherness of the same ontological status
> as the same, without there being immediately mounted an effort at its ap-
> propriation. . . . it is clear that the hegemonic impulse thematized in the
> chivalric quest was a fact of culture and that its failure in the political
> realm—witness the case of the wars upon Islam (the Crusades)—in no way
> invalidated its hold in other areas, especially in the practice of knowledge, as
> Edward Said has convincingly shown. Politically, the West may have had to
> grudgingly accept the existence of the Islamic otherness, but in the realm of
> knowledge it acknowledged no such possibility.[6]

As theorists of otherness, both Fabian and Godzich agree that for medieval culture there existed a conceptual hegemony—either religious or political (or both)—whereby nothing that exists could not be brought back into the fold of the dominant ideology. In other words, they argue that there can be no permanent alien or ontological other in premodern culture. But is this true, or is this just another example of modernist totalizing of the Middle Ages?

In looking at exotics in performance I would turn instead to Steven Mullaney's fascinating discussion of "The Rehearsal of Cultures," the third chapter in his study of the place of the stage in Renaissance England.[7] Mullaney analyzes the appearance of the *Wunderkammer*, or wonder-cabi-net, "a form of collection peculiar to the late Renaissance, characterized primarily by its encyclopedic appetite for the marvelous or the strange."[8] In contrast to the scientific museums that were organized in the seventeenth century, the wonder-cabinet is a "room of wonder, not of inquiry," featuring alien objects that are allowed to remain as "spectacles of strangeness," not to be subsumed by ordered systems of categorization. Mullaney wants to argue that this curiosity—the will to maintain the other for purposes of display and the cultivation of wonder—typifies late Renaissance culture. For Mullaney, the wonder-cabinet is emblematic of a concern with the multiplicity of cultural others that preoccupied the "dominant cultures of early modern Europe."[9]

Mullaney uses the concept of a "rehearsal of cultures" to describe a "period of free-play during which alternatives can be staged, unfamiliar roles tried out, the range of one's power to convince or persuade explored with some license."[10] He proposes that, like the wonder-cabinet, the Elizabethan popular stage served to display the other for purposes of cultural reflection. I'd like to extend his proposition back into medieval performance and across the channel to sixteenth-century Continental plays as well, for a fresh examination of relevant texts and records may suggest that a number of performative sites and occasions functioned to provide this display of the exotic other for purposes of aesthetic pleasure and cultural reflection during the fifteenth and sixteenth centuries.

Although drama scholars have occasionally treated the representations of Old Testament patriarchs in European biblical history plays as potential problems, they have generally used typological models to contain the patriarchs' difference or otherness and to produce identity and similarity in relation to New Testament portions of the narrative (which are always regarded as ideologically dominant). But what if the producers of the Old Testament plays intended a sustained portrayal of alterity? The detailed costume lists surviving from a 1583 Lucerne performance provide an entry into this possibility, for they clearly represent the patriarchs and prophets as both strange and exotic.[11]

In describing the effects to be produced by the costumes, the language of the 1583 costume list emphasizes the importance of strangeness. Jacob the Patriarch, for example:

> should be dressed in a somewhat different manner from the prophets, as an old patriarch, yet honorable and wealthy, in a quite old fashioned way, *the stranger the better* [emphasis mine, here and in the following examples] ("uff gar allte manier, ye seltzamer ye bessar"). The remaining prophets are to be dressed as prophets, *almost identical in a strange way* ("vast glychförmig uff seltzame manier") as far as the over-capes and hats are concerned; underneath long priests' robes, at the waist gathered in with broad belts or other strips. Each is to have a book and a staff in his hand, likewise boots, also long gray hair and a beard.[12]

The strangeness in Jacob's costume is connected to the "old fashioned" style; here temporal distance is understood to produce a dramatically attractive sense of alienation.

Within the category of the strange there can be distinctions, including class, for despite his deliberately heightened strangeness, Jacob is to be shown as "honorable and wealthy." Similarly, in the same costume list, Abraham "is to be dressed as a rich old patriarch in an expensive and strange style, *the stranger the more worth looking at* ("sol belkleidt syn alls ein Rycher allter Patriarch uff kosliche allte und seltzame Manier, ye seltzamer ye ansichtiger"), with boots and a saber at his side."[13] Young Isaac his son and the two servants who accompany them in the sacrifice-of-Isaac episode are to be dressed in a strange old fashion too, Isaac more richly than the servants. The Lucerne list thus prescribes a style that visually communicates "otherness," and clearly intends this feature of the performance to hold the audience's attention—the stranger the more worth looking at. Moreover, it is a nuanced category, for within it status and socioeconomic distinctions remain. Later Esau and Jacob will be dressed "in an old-fashioned and strange manner" that is also *expensive*—as opposed to the indigent, old-fashioned other, Joram, farther down on the list, who is dressed "as a vagabond in a poor Jewish style."[14]

Jewish dress was another possible category of the exotic in medieval and early modern culture, though in the case of the young Isaac it is invoked only to be rejected. "Isaac is to have a coat of blue with wide sleeves, open at the front, with buttons the same as the material, large, lined, reaching down to his knees, belted at the waist, with a yellow or green silk strip, a fine hat, *but not with a hood like the Jews,* and then boots to his calves but not pointed, yellow. The two boys can also be dressed like this, but in a different color."[15] The youthful Isaac and his servants are strange and exotic in a non-Jewish way. Thus we have several subcategories of the other within the semiotic system of the 1583 costume list.

The costumes and movements prescribed for the elderly Isaac and Rebecca his wife, however, suggest their Jewish otherness. Rebecca "is to be dressed in an honorable, costly, and rich manner, but *in the Jewish and old-fashioned style,* also with proud arrogant gestures."[16] The reference to "proud arrogant gestures" is a reminder of the long performance history of Jewish stereotyping. Stage directions in the Latin church dramas of the twelfth century indicate the stylized presentation of Jewish figures. Archisynagogus in the Benediktbeuern Christmas Play responds to prophecies of Christ's birth with "an excessive clamor; and shoving forward his comrade, agitating his head and his entire body, striking the ground with his foot, and imitating with his scepter the mannerisms of a Jew in all ways" ("imitando gestus Judaei in omnibus").[17] Within the strong typological frame of the Benediktbeuern play, the portrayal of the arch-Jew, Archisynagogus, is unambiguous. He represents the misguided intellectual/religious tradition that must be subordinated to the new revelation, and his violent but comic movements caricature the irrationality and "stubbornness" of which the Jews were accused in theological debates of the period. When we compare these twelfth-century stage directions with the late-sixteenth-century costume list, we must raise questions not only about the longevity of such Jewish stereotyping but also about the uniquely ambiguous or contradictory place of Old Testament figures in civic biblical drama.

The distinctions that the Lucerne list makes *within* this "old fashioned" and "strange" style of Old Testament costuming suggest that the category of other was not a monolithic one. As we have seen above, costumes distinguish socioeconomic status and degrees of Jewish identification although all characters are historically other. Even within the Jewish subcategory, distinctions can be made. Jethro, for example, is to be dressed in "a long, clean, honorable, and almost priestly clothing, but different from Moses and Aaron, yet Jewish."[18] The Jews who accompany Moses are "all to be dressed in good Jewish style."[19] This style is spelled out in specific detail: "In the procession of Israelites with Moses, all of them, both young and old, are to have brass rings or clasps hanging from their ears, and further they are to wear all the time on both days short boots in the Jewish style, yellow or blue."[20] Moses himself, however, is dressed "in a soldierly manner, something different, between a patriarch and a prophet."[21] The common "Jewish style" is thus subtly differentiated from the mode of dress of Old Testament patriarchs, prophets, or soldiers.

Non-Jewish categories of otherness also appear in the costume list from Lucerne. Putiphar and Sother, the Ismaelite or Egyptian merchants "are to be dressed in long robes with curved sabers, also tall hats with feathers, veils, as Arab (*Heydnische*) or Turkish merchants," while the Goldsmith is to be dressed "in Jewish fashion."[22] Like the non-Jewish merchants, the Goldsmith combines a trade category of costume with an ethnic category; what is notable is that being Jewish is here simply one ethnic identity among many possibilities. The Lucerne performance thus deliberately multiplies the categories of cultural and ethnic difference. These are combined with the fantastic in such figures as the giant Goliath, who is described as "got up and armed in a strange and heathen manner, in full armor with shield and stave. He is also to have a dummy head."[23]

Finally, the play distinguishes between Old and New Testament history through its costuming. Laban is dressed "like an old Jewish priest, but in a less costly fashion than Zacharias," who "is to have the same clothing as Aaron until the Old Testament is ended, for at the beginning in the procession of Israelites he is dressed like any other Jew. But when the New Testament starts he is a priest and the father of John the Baptist, keeping the same clothing until he comes to John the Baptist asking who he is; thereafter he puts on again his original clothing, which he retains to the end."[24]

Stella Mary Newton in her fascinating study of *Renaissance Theater Costume and the Sense of the Historic Past* notes that:

[a]rtists of the later middle ages thought of the New Testament as introducing the modern world—the present way of life—so that the saints, unless they were the Apostles, could reasonably be represented as wearing the kind of dress that would be worn by those in the modern world who corresponded to them in occupation or social position. Pilate and the High Priests of the Synagogue on the other hand could not, since they represented the old point of view, the world before the Redemption. They had, in addition, to be distinguished from each other and, as pictorial subject-matter was enlarged to include episodes from the Old Testament . . . there was evidently a feeling that these too must be shown by their dress as events earlier than those related in the Gospels or by St. Paul. The appearance of the Magi, who represented the first knowledge of the coming of Christ to those outside the Jewish world (as well as outside the Roman Empire), also had to be taken into consideration where costume was concerned. Oriental dress played, therefore, an important part in establishing those differences.[25]

Newton points out that fashions of about fifty years earlier were often used to convey the sense of the old-fashioned and historically distant in the visual arts and performance, while hats identified Jews or Orientals.[26]

The differentiation of antique costume in the lists—between Jewish and non-Jewish, Western and Oriental, heathen and Christian, and even between the subtypes of Jewish—visually reinforces our sense that we are shifting categories from the Old to the New Law or the West to the East, but it more importantly reveals that the categories of otherness are multiple, not monolithic. The historically or geographically distant is represented in carefully delineated variety that cannot be understood simply in negative terms.[27]

Similarly, the music sung by the chorus in the synagogue, the *Judengesange*, for the 1583 Lucerne performance requires an interpretation beyond the hostile representation of Jews. The music is still extant on seven music blocks (parchment stretched over wooden forms and bound in book form). Blakemore Evans, the modern editor of the stage directions, says: "The texts of these songs are composed of a strange mixture of sense and nonsense. They were the pride and joy of their composer Cysat [the producer-director of the 1583 performance], who claims to have used odds and ends of eighteen different languages in their composition. . . . Many of the nonsense words used in the choruses are undoubtedly taken from children's rhymes; many of them are still current to the present day." Blakemore Evans sees a striking similarity also "between many of the words in these verses and commonly current magic incantations."[28] The otherness of the Lucerne synagogue music is apparently not meant to provoke condemnation but rather to perform the fascination of the exotic.

The Old Testament characters featured in the costume lists are distanced by history and by ethnic or religious affiliation from the sixteenth-century audience of the Lucerne plays, but their otherness does not appear to be predicated on moral difference. They are neither totally assimilated to the audience identification group nor rejected as totally alien to it. They are, in a word, "exotics"—performing the alien in all its attractive and dangerous fascination. Anthropologist Stephen William Foster argues that the exotic, which "immediately evokes a symbolic world of infinite complexity, surprise, color, manifold variety and richness," is (to use Michel Foucault's term) an *episteme:* "a relatively fixed cultural problematic which becomes operational as an internalized *gestalt* and structures discursive activities pertaining to cultural difference."[29] It is this episteme, I would argue, that is visible in late medieval and early modern performance when historically or geographically distant figures take the stage.

As Foster theorizes the exotic, this episteme provides a cultural mechanism for comprehending remote and unknown phenomena without totally emptying them of their strangeness. Like the term "culture," the "exotic" labels and initiates control of social phenomena that, "being at the outset remote and unknown, may for those reasons appear chaotic, threatening, bizarre, ineluctable." Once such a label is imposed:

the phenomena to which they apply then begin to be structured in a way which makes them comprehensible and possibly predictable, if predictably defiant of total familiarity. The exotic is always full of surprises; it delights and titillates. To domesticate it exhaustively would neutralize this aspect of its meaning and regretfully integrate it into the humdrum of everyday routines. The ideology implied by the exotic therefore stops short of an exhaustive interpretation.[30]

The exotic thus remains a paradoxical category, representing both known and unknown simultaneously. It is "liminal," in Victor Turner's theoretical language.[31] Such a liminal category or symbolic mediator can never be totally one thing or another but remains, paradoxically, both things at once. Thus, to return to the two theorists with whom we began this analysis—Johannes Fabian and Wlad Godzich—medieval political or religious ideologies could never totally assimilate the exotic other because its very definition within social discourse requires that it retain a modicum of difference.

The Old Testament patriarchs and prophets are both us and not-us in the discourse of medieval biblical drama. Costuming them as exotics from an antique culture, portraying them as old-fashioned and alien, allows the producers of the drama to communicate their differences from medieval or early modern European Christian audiences while not stigmatizing them as social or spiritual outsiders. It is true that the Harrowing of Hell sequences dramatize the eventual recuperation of many of these pre-Christian types by Christ, so that from the perspective of cosmic history their similarities ultimately outweigh their differences. Nevertheless, I would argue that, as dramatized during the Old Testament sequences, the patriarchs and prophets are exotic figures whose visual otherness would surely counterbalance the audience's knowledge of their role within the narrative of Christian history, as the Lucerne costume direction makes clear when it says "the stranger the more worth looking at." Given the construction of exotic types like the Jew or Oriental through stereotypical costume and gesture, and given the ambiguity inherent in the category itself, it is clear why the content of exotic types was unstable. Because of this instability, the paradoxes of the exotic tended to break down under the pressure of a dichotomous moral system of good and evil—like that of the New Testament plays, where Jews and pagan tyrants are almost always bad characters.

In the New Testament sequences, the role of nonstigmatized exotic other is most often played by the magi, Oriental kings who nevertheless traverse the gap between places and cultures to worship the infant Christ. In Matthew, the wise men who come from afar are referred to as "magi," or

Persian priests. Their exoticism is best seen in the visual arts of the medieval and early modern periods. They are given exotic costumes as early as A.D. 432 in the Roman church of Santa Maria Maggiore,[32] and by the twelfth century were always portrayed as kings from exotic Eastern realms, which included Africa. John of Hildesheim's *Historia Trium Regum* (ca. 1375) calls one of the three an Ethiopian, and most fifteenth-century paintings include a black magus. Records of an Italian confraternity describing a 1497 Corpus Christi procession note not only that the Old Testament characters are dressed in Jewish costume ("vestito ala Zudea") but identify one of the magi as black.[33] The Lucerne costume list, not surprisingly, calls for Balthasar to be a Moorish king, exotically clothed and riding a camel.[34]

Paul Kaplan, who has exhaustively surveyed *The Rise of the Black Magus in Western Art,* sees the story of the magi as the "preeminent means of integrating the inhabitants of the non-European world into the western Christian universe." He notes with some puzzlement the unstable and contradictory associations blackness could have, even in the context of a strongly positive religious myth.[35] However, given the function of the exotic in mediating contradictions, such "unstable and contradictory" associations should not be surprising. Neither is it surprising that the black magus became a fixture in Epiphany painting of the sixteenth and seventeenth centuries—the era of world exploration, colonization, and the rise of a global slave trade. Richard Trexler has argued that representations of the magi in Italian art were available to legitimize the new world of conquest,[36] but I would suggest that we not leap to the ultimate political function before exploring the mediating aesthetic and cognitive functions of such images. The major purpose of the exotic as interpretive mechanism is, in Stephen Foster's view, to allow "members of one social group to understand another social group that they see as different from their own." The exotic operates as a symbol that combines cultural differences, even when they are seen as opposites, "in the interest of traversing boundaries between cultures."[37] An anthropological model such as Foster's, while not ignoring power differences between cultures, is willing to allow more interpretive space to the experiencing of that difference. Renata Wasserman, too, in her analysis of the language of exoticism, points out that the exotic is an "acceptable guise," a "mediating entity," by which nations could examine and critique a variety of basic but evolving assumptions about identity.[38] Such models call attention to the pleasure-making and the cognitive functions of the exotic that Renaissance scholars have recently associated with "wonder," as a phenomenon in the development of science, exploration, and theatre.[39]

As Hugh Honour points out in his survey of European images of America from the time of the discoveries, "In a Portuguese painting of the

Adoration of the Magi probably dating from the first decade of the sixteenth century, the place traditionally occupied by the black Magus has
been taken by a coppery-skinned Brazilian" wearing gold jewelry, a feather
headdress, and a richly patterned shirt.[40] The Brazilian is portrayed as alien,
but with good intentions approaching a very European Christ child. As
Honour notes, the discovery of Brazil in 1500 initiated a century-long de-
bate on the spiritual status of the Indian, which can be read in the wide
spectrum of Indian images from the early modern period. Just as with the
Jewish or black exotic, the Indian exotic participates in an unstable category that attempts to balance the alien with the benign.

This balancing act is accomplished primarily by transforming the exotic into an *objet d'art,* a focus of wonder and contemplation. Among the
varieties of exoticism to be found at court festivals and royal entries during the fifteenth and sixteenth centuries was the Turk. Continuing tensions
between Christians and the Infidel through the period might suggest a direct political significance, but the Turk is rarely portrayed as a villain.
Rather, Turkish figures are colorful and even jolly participants in pageants,
tournaments, and processions in England and on the Continent. Scholars
of these entertainments have puzzled over the portrayal of lighthearted exoticism where they expected demonizing, but anthropological theories of
the exotic emphasize that the ideological bridge between the familiar and
the alien is aesthetic.[41] We should therefore expect performance in this period to be filled with types of the exotic: Oriental kings, Jews, Turks, even
wildmen.[42]

In courtly spectacle, the exotic is rarely problematic. At banquets or
masques, for example, where wildmen suddenly appear or exotics are displayed, the interplay between known ritual and exotic surprise is part of
the performance code. Certainly with a small number of participants of the
same class in a restricted private setting—where *sameness* is the norm and
difference can be interpretively controlled—the strange and surprising appearance of the exotic adds an almost risk-free piquancy to the event.

The more interesting performance of the exotic occurs in public performances with a mixed audience of considerable size, where the alien is
subject to fewer controls (either spatial or interpretive) and is therefore
likely to be a more volatile element. Civic festivals often enact legislation
designed to control both the playing of and the audience response to the
other—whether that is the demonic, the miraculous, or the exotic.

As a semiotic category, the exotic is triggered by visual signs like costume and gesture. The ephemeral nature of these aspects of the theatrical
code have made it difficult for medieval scholars to perceive exotic elements in characterization, so that the visual arts of the late Middle Ages
and the early modern period offer the clearest proof of how ubiquitous

exotic images were. Furthermore, the dialogue of surviving play texts often does not participate in the exotic representation. It may even trigger an alternative interpretation, so that the experience of watching biblical history plays was likely one of semiotic overload through conflicting and competing codes rather than the one of systematic and totalizing control cultural critics hypothesize.

The Towneley Old Testament plays, for example, are structured by charismatic patriarchs and prophets in dialogue with the divine will, and in striking opposition to dominant evil characters like Lucifer, Cain, Pharaoh, etc. More than any other English cycle, the import of such characterization seems to be to set up a line of good and obedient servants of God who will offer contrast to a line of evil, disobedient rulers. This produces a dualistic political system that would seem to make any representation of the "exotic other" almost impossible, given a definition of the exotic as a category that mediates difference. We have no surviving evidence for costume and gesture in any production of these plays, but it is thought-provoking to imagine the code clash that would occur if such good patriarch and prophet figures were dressed as the Lucerne costume list suggests.

Bringing the exotic other into the Old Testament sequence through costuming would complicate the "us-them" affiliations the scripts set up. What happens when "we" (the laudable group) are simultaneously "old fashioned" and "strange" (to use the Lucerne terms)? Would the visual trigger for the exotic tend to override the verbal systems of meaning? Such questions, to which firm answers are mostly not available, point toward a reading of the theatrical experience of civic biblical plays that is far less reductive than that usually offered by textual scholars.

Notes

1. "Liminality, Carnival and Social Structure: The Case of Late Medieval Biblical Drama," in *Victor Turner and the Construction of Cultural Criticism: Between Literature and Anthropology,* ed. Kathleen M. Ashley (Bloomington: Indiana University Press, 1990), 42.
2. This is the interpretation popularized by Edward Said in *Orientalism* (1978; reprint. New York: Random House, 1985). On "Orientalism," "Other/other," and "Othering," see Bill Ashcroft, Gareth Griffiths, and Helen Tiffin, *Key Concepts in Post-Colonial Studies* (London: Routledge, 1998), 167–173. See also the essays collected in the volume *"Race," Writing, and Difference,* ed. Henry Louis Gates, Jr. (Chicago: University of Chicago Press, 1986).
3. Peter Mason, *Deconstructing America: Representations of the Other* (London: Routledge, 1990), 43. For a similar argument—that the tragedians of ancient

Greece used the "barberoi" to define their own Athenian culture—see Edith
Hall, *Inventing the Barbarian: Greek Self-Definition Through Tragedy* (Oxford:
Clarendon Press, 1989).

4. Peter Mason's chapter on "Popular Culture and the Internal Other" de-
    scribes the available others of medieval Europe, especially the wildman and
    -woman, whose iconography was available for projection onto the others
    discovered in the New World; *Deconstructing America,* 41–68. He cites P.
    Vandenbroeck's study *Beeld van de Andere, Vertoog over het Zelf* (Antwerp:
    Royal Museum for Fine Arts, 1987).

5. Johannes Fabian, *Time and the Other: How Anthropology Makes Its Object*
    (New York: Columbia University Press, 1983), 26.

6. Wlad Godzich, "Foreword: The Further Possibility of Knowledge," in
    Michel de Certeau, *Heterologies: Discourse on the Other,* trans. Brian Massumi
    (Minneapolis: University of Minnesota Press, 1986), xiii.

7. *The Place of the Stage: License, Play, and Power in Renaissance England*
    (Chicago: University of Chicago Press, 1988), 60–87.

8. Ibid., 60–61.

9. Ibid., 63.

10. Ibid., 69.

11. Costume lists for the First Day of the 1583 performance, with additional
    entries from 1583 and 1597, are translated in *The Staging of Religious Drama
    in Europe in the Later Middle Ages: Texts and Documents in English Translation,*
    ed. Peter Meredith and John E. Tailby, EDAM Monograph Series 4 (Kala-
    mazoo: Medieval Institute Publications, 1983).

12. Meredith and Tailby, *Staging of Religious Drama,* 131. The originals are
    found in M. Blakemore Evans, *The Passion Play at Lucerne* (New York:
    MLA, 1943), 193. See also the discussion by Hansjürgen Linke, "Germany
    and German-speaking central Europe" in *The Theatre of Medieval Europe:
    New Research in Early Drama,* ed. Eckhard Simon (Cambridge: Cambridge
    University Press, 1991), 207–224. Linke notes that "Lucerners always called
    their great medieval play 'Easter play,' but to our way of classifying—since
    it contains both Old Testament episodes and the Crucifixion, all presented
    over two days—Lucerne had a 'Passion play.'"

13. Meredith and Tailby, *Staging of Religious Drama,* 133; Blakemore Evans, *Pas-
    sion Play,* 194.

14. Meredith and Tailby, *Staging of Religious Drama,* 134.

15. Ibid., 133.

16. Ibid.

17. Archisynagogus' stage behavior may be found in *Medieval Drama,* ed. David
    Bevington (Boston: Houghton Mifflin, 1975), 183.

18. Meredith and Tailby, *Staging of Religious Drama,* 135.

19. Ibid.

20. Ibid., 136.

21. Ibid., 135.

22. Ibid. Various costumes featuring exotic turbans, hats with feathers, and curved sabers may be found in Cesare Vecellio's *Habiti antichi, et moderni di tutto il Mondo* (Venice: Giovanni Bernardo Sessa, 1598), reproduced as *Renaissance Costume Book* (New York: Dover Publications, 1977). Vecellio was a Venetian who assembled a book of 420 woodcuts, of which 361 focused on European and Turkish costume, while the other fifty-nine represented Asian and African costume.

23. Meredith and Tailby, *Staging of Religious Drama*, 137.

24. Ibid., 136.

25. Stella Mary Newton, *Renaissance Theater Costume and the Sense of the Historic Past* (London: Rapp and Whiting, 1975), 70–71.

26. Newton, *Renaissance Theater Costume*, 90. The prophet Zacharias at Claus Sluter's Puits de Moise sculpture of the prophets (before 1406) near Dijon, France, wears a large, slouchy Jewish-style hat, and Newton even speculates that the dress, beards, and scrolls held by this sculptural group are so theatrical that they must have been copied from actors wearing stage costume (90–94). See also Jacques Thibaut's description of the staging of the *Mystère des SS Actes des Apostres* at Bourges in 1536, with many characters dressed "à la mode antique" (222–26).

27. The negative effect of Jewish costume was doubtless intended in the Passion of Donaueschingen, where a symbolic figure Judea, who is disputing with "Christiana," is said to be "judaiquement vêtue"; from Gustave Cohen, *Histoire de la Mise en Scène dans le Théâtre Religieux français du Moyen Age* (Paris: Champion, 1951), 223. See, however, the fascinating discussion of the utility of racial difference for "exploring the dialectics of difference and sameness within an expanding series of contexts" offered by Claire Sponsler and Robert L. A. Clark in "Othered Bodies: Racial Cross-Dressing in the *Mistère de la Sainte Hostie* and the Croxton *Play of the Sacrament*," *Journal of Medieval and Early Modern Studies* 29 (1999): 61–87.

28. Blakemore Evans, *Passion Play*, 68.

29. Stephen William Foster, "The Exotic as a Symbolic System," *Dialectical Anthropology* 7 (1982): 21–30.

30. Ibid., 21–22. For a contemporary reading of the souvenir as exotic object that is appropriated to private space, see Susan Stewart, *On Longing: Narratives of the Miniature, the Gigantic, the Souvenir, the Collection* (Baltimore: Johns Hopkins University Press, 1984), 132–166.

31. On Turner's theories of "liminality," see Kathleen M. Ashley, ed., "Introduction," *Victor Turner and the Construction of Cultural Criticism: Between Literature and Anthropology* (Bloomington: Indiana University Press, 1990), xviii–xix.

32. See Andre Grabar, *Early Christian Art: From the Rise of Christianity to the Death of Theodosius* (New York: Odyssey Press, 1968).

33. G. Mantese, "Congregation ad honorem sacratissimi," in *Rivista di storia della chiesa in Italia* 15 (1961): 109–22. I would like to thank my colleague Ellen Schiferl for bringing this record to my attention.

34. Blakemore Evans, *Passion Play*, 201.
35. *The Rise of the Black Magus in Western Art* (Ann Arbor: UMI Research Press, 1987), 105. A catalogue of representations of the magi in the art of Cologne, whose cathedral houses the relics of the Three Magi, is *Die Heiligen Drei Konige: Darstellung und Verehung*, Katalog zur Ausstellung des Wallraf-Richartz Museums in der Joseph-Haubrich-Kunsthalle (Koln, 1982).
36. Richard Trexler, *Bearing Gifts: The Magi Cult and the Documentation of Social Processes* (Binghamton: Fernand Braudel Center, 1980).
37. Foster, "The Exotic as a Symbolic System," 22.
38. Renata R. Mautner Wasserman, *Exotic Nations: Literature and Cultural Identity in the United States and Brazil, 1830–1930* (Ithaca: Cornell University Press, 1994), esp. 13–33, 244–259. Wasserman says her project is "to show that interaction can have results other than isolation or destruction" (33). Another recent theoretically sophisticated rejoinder to Said's "Orientalist juggernaut" is Lisa Lowe's *Critical Terrains: French and British Orientalisms* (Ithaca: Cornell University Press, 1991).
39. A delightful survey is Lorraine Daston and Katharine Park, *Wonders and the Order of Nature, 1150–1750* (New York: Zone Books, 1998). In chapter two, "The Properties of Things," they comment on the courtly preoccupation with wonders in both nature and art as forms of symbolic power. The dukes of Burgundy, in particular, were attracted to the exotic and the strange, manipulating the emotion of wonder and the discourse of the marvelous to communicate their wealth and their Eastern political and military goals (100–108). A persuasive study of the philosophical backgrounds to Shakespearean plays like *Pericles*, Cymbeline, *The Winter's Tale*, and *The Tempest* may be found in Peter G. Platt, *Reason Diminished: Shakespeare and the Marvelous* (Lincoln: University of Nebraska Press, 1997). Platt includes a chapter on "The Masque and the Marvelous" that takes on the dominant interpretation of the genre—that the masque is all about containment (99–123). Platt argues for an alternate theory of wonder, articulated by Francesco Patrizi and demonstrated by Shakespeare, that subverts and destabilizes the role of reason (18). See also T. G. Bishop, *Shakespeare and the Theatre of Wonder* (Cambridge: Cambridge University Press, 1996).
40. Hugh Honour, *The New Golden Land: European Images of America from the Discoveries to the Present Time* (New York: Pantheon Books, 1975), 53.
41. For puzzled scholarly reactions to representations of the Turk, see Roy Strong, *Art and Power: Renaissance Festivals 1450–1650* (Woodbridge, Suffolk: 1984), 135; also Clarence D. Rouillard, *The Turk in French History, Thought and Literature (1520–1660)* (Paris: Boivin, 1941). Rouillard notes the Turks in the 1468 Bruges tournament where, instead of representing infidels, they added an exotic and spectacular element.
42. The wildmen of medieval and Renaissance art and spectacle, especially the courtly masque, provided the same frisson of mingled danger and pleasure. See examples cited in Sydney Anglo, *Spectacle, Pageantry, and Early Tudor Policy* (Oxford: Clarendon Press, 1969), 119, 122, 247. Barbara Ketcham

Wheaton argues that in the late Middle Ages three new elements were added to the ceremonial feasts of the aristocracy: disguising, mumming, and interludes. The motif of the surprising and unexpected arrival at the feast of some exotic figure like a fantastically dressed Saracen, a wildman, an Indian, or exotic animals like elephants, dromedaries, or lions became part of the play on the expected/unexpected for these events; *Savouring the Past: The French Kitchen and Table from 1300 to 1789* (London: Chatto & Windus, 1983), 8, 21.

Chapter 5

On Making Things Korean:
Western Drama and Local Tradition in
Yi Man-hûi's *Please Turn Out the Lights*

Jinhee Kim

It has become commonplace to assert that Western forms and themes
have massively invaded non-Western values and worldviews. Surprisingly often in postcolonial as well as colonial criticism, the West is portrayed as the perpetrator, while the non-Western population is cast as the
victim, once subjugated by Western political imperialism and now dominated by a subtle yet equally destructive cultural imperialism. It is not uncommon to hear postcolonial scholars voicing fear that Western hegemony
will eventually obliterate indigenous cultural identities and their will to
power.[1] From this perspective, the hegemony of the West is so extensive
that it jeopardizes the very existence of non-Western cultures.

Clearly such a view is, at least in part, justifiable. Yet it hardly represents
the whole truth about contemporary intercultural communication. Indeed, claims about processes of Westernization, official or not, in non-
Western cultures deserve a more thorough investigation. In particular, the
concerns raised by postcolonialists tend to miss a point that is indispens-
able to understanding cross-cultural dynamics. Contrary to the pervasive
perception that the growth of Western models of literature, religion, in-
dustry, leisure, and politics dominates indigenous populations, a look at
specific instances shows that cultural exchanges can be astonishingly diverse and complex. Furthermore, closer investigations of cultural interaction often reveal that non-Western populations are not particularly
hampered by the disadvantages so often theorized in discourses of the
hegemony of the West.

In this essay, I refute the notion that the ready acceptance of Western
literary models inevitably implies the wholesale imposition of Western
modes of thought, thus eradicating indigenous traditions and ultimately
the cultural identity of non-Western audiences. I argue instead that the
adoption of Western models does not deprive indigenous populations of
their cultural roots, especially since non-Western audiences are not, and
cannot be, mere passive receivers of hegemonic Western texts and their
accompanying ideologies. The ways in which Korean playwrights and
theatre companies have appropriated material from Western drama, as
well as the manner in which Western-style Korean drama has been received, indicate that Korean writers, directors, and audiences apply their
own experience and the conventional values of their own cultural tradition to their understanding of Western and Western-style theatre. In fact,
Western-style Korean drama suggests that the allegedly devastating effects of Western models are themselves more a Western construction than
a Korean reality.

My main concern in this study is with the reception of Western-style
drama performed in Korea for Korean audiences. Modern Korean drama
offers a particularly instructive case study, because in Korea the modern
theatre is considered fundamentally Western in its theories and practice.
Unlike traditional types of performance such as the Mask Dance, modern
Korean theatre reproduces Western realist theatre in all its formal aspects,
from the plots and style of the playscripts to staging, scenery, and acting.
Whereas traditional performances require open arenas and employ acting
techniques that rely on spontaneity and improvisation, modern Korean
drama involves a formal separation of performer and audience, of stage and
auditorium. Basing itself on realistic modes of representation developed in
nineteenth-century European theatre, contemporary Korean theatre is designed essentially to manifest what is supposed to be private (feelings, situations, conflicts), while concealing the very artifice of such manifestation
(machinery, painting, makeup, light sources, music, etc.). This apparent assimilation of a foreign mode of representation has led Korean audiences to
think of modern Korean drama as preeminently a Western import. As a
consequence, Western-style drama has long been denied critical attention
as well as acclaim, and has by and large been perceived as incapable of nur-
turing indigenous cultural values and as alienating Koreans from their own
cultural heritage.

While some Koreans have echoed Western postcolonial scholars in
claiming that the adoption of a Western dramatic style has been a form of
cultural imperialism,[2] Korean audiences are capable of responding, within
the framework of the traditional values of their culture, to plays based on
Western models performed in Western-style productions. A particularly

good example of such sophisticated audience response is a recent Korean play by Yi Man-hûi, *Please Turn Out the Lights.*

Yi was commissioned to write this drama to celebrate the sixth anniversary of its host theatre, the Taehakro Kûkchang (the College Street Theatre) in Seoul. Premiering on New Year's Day 1992, Yi's play was immediately acclaimed by a number of daily newspapers as a new kind of drama, one that shrewdly investigates the dilemmas of late-twentieth-century humans caught—particularly in ever-expanding and ever-Westernizing Korea—between personal desires and social obligations. Subsequent reviews echoed these claims. The critical success of the drama was accompanied by enormous commercial returns. Each ticket cost W10,000, approximately $12.50. Of the twelve plays participating in the sixteenth Seoul Theater Festival of 1992 in the category of Free Competition, *Please Turn Out the Lights* commanded the second highest ticket price along with three other plays.[3]

The production management immediately decided to extend the run of the play to a full year and beyond. Kang Yông-gôl, the director of the play, noted that "our initial plan was to run the play for three months and then decide whether or not to extend the run of the play another three months, and possibly to a full year. After one month's production, we decided to extend the run for another five months. After our third month, we decided to push the run to a full year."[4] The enormous popularity of the drama resulted in unusual longevity for a theatrical production in Korea: on February 27, 1994, *Please Turn Out the Lights* celebrated its one-thousandth performance, an outstanding record for an original Korean play, and it was estimated to have attracted more than 150,000 viewers up to that time.[5]

The playwright Yi noted on the original program book in 1992 that "the focus of *Please Turn Out the Lights* is not on romance but on revealing the duality of our lives—our divided selves and the pains arising from the gap created by the two conflicting paradigms of modern life."[6] What he wants to tell us is not a love story but a testimonial to our life, the story of the struggle of a modern man in quest for his true identity. Yi is convinced that his play has more to offer than just its innovations in form and style. Yi sees himself as a social critic whose mission is to portray candidly the contemporary Korean situation. There is a sense of warning and even reproach in his assessment of the Korean situation: that those who strive to fulfill their dreams must anticipate a conflict between their own personal desires and the built-in goals of the social institutions in which they operate. In Yi's view, the conflict not only delays personal gratification but also produces vice rather than virtue. Considering that Yi identifies himself as a playwright who writes about Koreans for Koreans, it is not surprising that he intends his artistic endeavors as positive contributions to the historical time and place he inhabits.

Perhaps an especially significant indication of the nature of the creative climate surrounding *Please Turn Out the Lights* is the fact that its performances have drawn large numbers of people to the theatre who do not regularly attend, especially middle-aged, male, white-collar workers. Kim Yun-ch'ôl, a drama critic and English professor at Sejong University in Seoul, notes that he was impressed by the turnout for *Please Turn Out the Lights* because middle-aged men who rarely treat themselves to dramatic performances made up about eighty percent of the audience.[7]

Please Turn Out the Lights is concerned with a middle-aged politician whose career is about to come to an end. The male protagonist, Kang Ch'ang-yông, visits his long-lost lover, Pak Chông-suk, on the eve of his decision to resign from his congressional seat. Kang is a career politician who rose to his present heights from the depths of poverty by marrying the daughter of a wealthy businessman. Kang reveals to Pak that his resignation from public office is motivated by remorse for his unethical conduct in the past. Until now he has completely concealed the fact that he killed a man who had frequently bullied and harassed the mother of a close friend. It was after he fled to a small town in the country and gained employment as a day laborer in an orchard that he met Pak, who was teaching art classes in the area. Though she was a respectable schoolteacher, who apparently did not know that he was a murderer on the run, Kang and Pak fell in love. However, Kang disappeared one day, feeling that he was not good enough for Pak and that his leaving would save her from future indignities.

Back in the city, Kang remained hidden, working at a construction site, again as a day laborer. It was there that he met his present wife, the daughter of Kang's employer, the chairman of the construction company. She insisted that despite Kang's humble background, they would make a happy couple. Aided by her fortune and her family's connections, Kang was able to gain a new identity and eventually become an important political figure. Now, ten years have passed since he was transformed into a political Cinderella; yet, as the drama begins, Kang is far from happy. His marriage is in trouble; his wife's automaton-like affections have become an ordeal rather than an inspiration for him. Her all-consuming sacrifices for the sake of his retarded son have come to bind Kang to a moral yoke. The very love and care displayed by his wife make Kang grow more unhappy day by day, because he recognizes that between him and his wife there is not "instinctive" love, but only "institutional" bonds.

Yet Kang is not the only one in crisis. For Pak, the past ten years have been as painful as they were for Kang, if not more so. The day that Kang disappeared, Pak set out to learn his whereabouts. After many false starts, she succeeded in locating one of her lover's close friends, Kim Tal-ho, and hoped through this connection to be reunited with Kang. However, Pak's search for her lost lover came to an abrupt end when she was betrayed and raped by Kim. The shame of her situation caused Pak to give up her search and marry her violator, because to her mind Korean culture offered virtually no alternative to, no escape from, the shameful situation imposed upon her. This decision seemed even more correct to Pak when she discovered that Kang had, for whatever reasons, married his present wife. Meanwhile, as might be expected from its inauspicious start, Pak's marriage has been nothing but a disaster. When she appears on stage, her bullying husband has long since left for America, leaving Pak behind in extremely poor health.

Most of the play is taken up with the gradual revelation of the past that I have just summarized. Yet there is more. Just when both characters seem to have completely exhausted themselves in making all these painful revelations, Kang makes an additional confession: his retarded "son" is, in fact, his half brother, born out of wedlock to his mother, a widow. Kang learned about his mother's pregnancy during one of the short breaks from his military service, but there was nothing he could do to amend, or undo, the situation. Kang goes on to admit that his mother, in shame, committed suicide immediately after this child was born, leaving him in Kang's custody. With this revelation, the play is brought to its denouement. The very last line of the play encapsulates the drama in its entirety in a terse but highly intriguing manner. In a soft but assured voice, the female protagonist, who has been listening to Kang, invites him to "turn out the lights."

At first glance, this text appears to be a very Western one. It can be most directly and readily placed within the heritage of Ibsen. Its structure is Ibsenesque in nearly every detail. Like Ibsen in *Ghosts*, where the narrative centers around the arduous process of revealing the past, the Korean playwright sets his play in motion through a series of confessions. Reading Yi's play is like peeling an onion. In *Please Turn Out the Lights* the importance of the revelations increases as the play progresses, for at each confession the audience is introduced to a more serious and devastating piece of truth. Whereas Kang and Pak take turns with their confessions in *Please Turn Out the Lights*, Ibsen establishes Mrs. Alving, the widow of Captain Alving, as the only character in a reasonable position to elucidate for the audience the problems that befall other characters. She has been able to conceal the truth from others for more than twenty years. But when her son, Osvald, returns to Norway as a grown man, she finds herself confronted with the reality of his hereditary illness and finally breaks her long silence. Osvald has been diagnosed as carrying a deadly disease that is about to leave him permanently paralyzed. Mrs. Alving's struggles to protect the reputation of

her family and to secure her son's future are echoed in Kang's actions, as
he covers up his own past and the sins of his mother. Mrs. Alving's original decision to remain a dutiful wife to her dissolute husband even after
his death and to see to it that the community remembers her husband as
a generous and kind person by dedicating an orphanage in her husband's
name are meant as a cover-up, to put a permanent seal on the secrets of
the Alving family. Yet this is not the conclusion of the revelations and realizations: Mrs. Alving comes to realize that it was she herself who drove
her husband to what she considered his dissolute lifestyle, because she did
not and could not respond to his needs, and that she as well as her husband were ultimately the victims as well as the agents of a closed, repressive, and hypocritical society. And the final question of the play is whether,
for once, she can step beyond what is "proper" and help her son out of his
misery.

In both plays, the consequences of the sins of the fathers are devastating. In *Please Turn Out the Lights,* the result is a retarded child, whereas in
Ghosts the fruit of Captain Alving's licentious lifestyle is Osvald's ruined
health. Yi employs the metaphor of deformity to portray the consequences
of sin. The illegitimate child is an "unnatural" product, and furthermore,
the child's physical inadequacy anticipates his inability to adapt to society.
A physical deficiency has a profound impact in *Ghosts* as well. Osvald was
sent off to Paris at seven by Mrs. Alving, who wanted her son to grow up
unaware of his father's "sin." But when Osvald returns to Norway, he is
crippled by his illness. Ibsen makes it clear that the sins of the dead disre-
gard distance; children have no choice but to live with the sins of their fathers. Ibsen's audience is in the end invited to understand Captain Alving,
to see him more as a victim than as a villain: our view of him turns around
twice! The final accusation in *Ghosts* is directed at a society that does not
allow the individual to unfold freely: the sinful "past" is not that of an individual but of an entire culture. This poses a point of contrast with *Please
Turn Out the Lights,* in which the audience is invited to accept the "sin" of
Kang's mother from the point of an "objective" value judgment, as something truly to be condemned and thus justly punished by the birth of a
handicapped child and her suicide.

The plot structure of *Please Turn Out the Lights* is perhaps its most Western feature, but it is one among many. For instance, the written stage directions spell out with great specificity the realistic details of the apartment as
displayed to the audience. There is a kitchenette on the right and a wooden
bookcase nearby. Across the room are a desk, a sofa, and a bed. A landscape
painting is hung on the wall. An easel and painting utensils, which indicate
the female protagonist's profession, are seen on the floor. This lifelike stage
is in some way reminiscent of the famous living room in Ibsen's *Nora,* since

in both plays stage props are employed to maintain a continuity between the theatrical illusion and the daily life of the spectators.

Just as much to the point, however, is the fact that the illusion of realism in this drama also conjures up images of 1990s Korea. The details of the setting of *Please Turn Out the Lights* position the play within an ordinary milieu, one encountered each day by middle-class Koreans; in fact, the play moves almost entirely within a space that strikingly resembles the space inhabited by most Koreans. What is striking about *Please Turn Out the Lights*, then, is the way in which its most pronounced Western features—derived from an Ibsenesque theatrical tradition—are not estranging aspects of the play but instead deeply familiarizing techniques. By invoking a Western style, Yi's play actually becomes more Korean.

The adoption of Western staging techniques is traceable to the single most significant historical fact about Korea in the first half of the twentieth century—the Japanese occupation. Among the many modernist projects engineered by the colonial Japanese government and the Japanese settlers between 1910 and 1945, the years of Korea's official colonization by Japan, was the construction of a proscenium-arch theatre, which transformed performance spaces. Before the construction of the proscenium-arch theatre, there was only one indoor performance tradition in Korea to which the commoners had access. Run by private subscriptions, it was known as P'ansori. P'ansori, a musical dramatization of narratives, its texts based on novels dating from previous centuries, was performed by a solo singer who was almost invariably male. Its six most popular pieces, which to this day are performed on a regular basis, are diverse both in tone and setting and in the seriousness of their themes. *Ch'unhyang'ga* [*Song of Ch'unhyang*] is a portrait of a young, beautiful woman born to a lower class, and of her undying love for a man whom she cannot legally marry. *Simch'ŏng'ga* [*Song of Simch'ŏng*] depicts a filial love in which the courage and sacrifice of a daughter prove powerful enough to win back her father's eyesight. *Hŭngbuga* [*Song of Hŭngbu*] is a story of poetic justice: the poor but good-hearted Hŭngbu receives a divine blessing, whereas his callous, greedy older brother loses his amassed wealth. *Sugung'ga* [*Song of the Underwater Palace*] spins the tale of a sea turtle and a white rabbit: the turtle invites the rabbit to the palace of the underwater world, while his eyes set on the rabbit's liver, which can save the ailing underwater king. *Paebichang t'aryŏng* [*Song of the Official Paebi*] is a lighthearted lampoon of a notorious government official with an excessive penchant for wine and women, while *Changhwa hongryŏngjŏn* [*Story of Two Sisters named Changhwa and Hongryŏng*] is a somber piece, meant to console the spirits of two sisters murdered by their own stepmother. All other identifiable forms of traditional Korean performing arts, such as T'alch'um [Mask Dance] and Nong'ak

[Farmer's Music], were, and in almost all cases continue to be, staged out-doors.[8] This was the traditional performance heritage that the proscenium arch transformed.

Another Western feature of *Please Turn Out the Lights* is its use of stage-setting conventions. When the stage is lit at the play's beginning, the center of a small apartment is seen. There is no curtain in the theatre. As a matter of fact, the theatre space is wide open. Approximately 150 seats surround the small stage, which is located at the bottom of the theatre. This stage structure, based upon the modern model, not only locates the audience close to the stage but also yields a close-range bird's-eye view from three different angles. In Western-style theatre, an audience that sees another part of the audience across the stage is, formally and perhaps psychologically, always being reminded that this is taking place in a theatre: "illusion" in its fullest sense is not possible. The stage as employed in *Please Turn Out the Lights* reinforces the established conventions of realism, and renders the audience the effect of empathy and identification.

Other aspects of the performance also point to the tradition of the Western realist theatre at work and thus paradoxically make the play seem a mirror of everyday modern Korean life: for example, the costumes. The male protagonist is dressed in a dark-colored tailored suit, a white shirt, a matching silk tie, and dress shoes. The formality of his attire is the norm in downtown Seoul, where each day millions of Koreans dress like that for work. Kang's suit is something that can easily be bought in a department store. This sense of familiarity is abundant in the outfit of the female protagonist as well. Her dress is silk, long enough to cover her knees, and its sleeves cover her arms. She is wearing a pair of earrings, a necklace, and a bracelet whose dark color and shape match her dress and hair. Her permed hair comes down to her shoulders and is firmly secured by hairpins. Her makeup is very ordinary in the sense that her face will never stand out in a crowd; in fact, her face is the sort that might remind an audience member of some neighbor.

While the actors' clothes, hairstyles, makeup, and accessories show a simplicity in taste and design typical of the lifestyle of middle-class Koreans, the onstage props further deepen the drama's essential sense of ordinariness. In front of the three-cushion fabric sofa, which is located in the center of the stage, a rectangular wooden table is seen. On it are placed a telephone, a Coca-Cola bottle, and a coffee mug.[9] It is not difficult to find this kind of furniture and paraphernalia in contemporary Korean households. Innumerable cues in the characters' appearances and in the stage setting connect the "world" created by the theatrical production to the real life of the audience, resulting in a striking aura of verisimilitude.

Another element that places Yi's play in the tradition of the European realist theatre while also linking it to modern Korean life is its dialogue. The manner of speech characters employ to express emotions and address one another is very much like that of Korean soap operas—another Western-inspired form. The dialogues between the protagonists are colloquial and idiomatic in vocabulary, replete with bits of common speech and contemporary vernacular Korean; slang and swearwords crop up frequently. The repeated employment of slang and trendy expressions can appear blunt and rather offensive to some audience members. But in general the intended audience, young college students and white-collar workers, seems to have had little trouble accepting the dialogue, and to have responded to it with amusement.

Finally, the production style of the play as it was presented in Seoul evokes Western theatre practice. As soon as the stage is lit at the play's opening, we see Kang and Pak in a tight embrace, their bodies leaning toward the stage corner nearest the audience. As if their physical closeness is something that sets off the tone of the drama, strong beams from several overhead lights catch each movement of these actors. Kang gently places his arm over Pak's shoulders, kissing her occasionally while his other hand is groping and feeling her body. When the steamy love scene, which lasts for several minutes, is over, the actors disengage from each other's arms and quietly walk together to the sofa. Here they sit in a rather detached manner, as if to negate the intimacy they have just displayed.

This sort of realistic staging and acting clearly points to Stanislavsky and his descendants, especially via their effect on contemporary television. Indeed, every stage action seems to have been designed both to "reveal" character in a Stanislavskian manner and to evoke as much emotional empathy from the theatre audience as possible. Encouraging the actor to approach the character to be represented as if he were a real person, even to the point of entering the character's unconscious, Stanislavsky assumed that the successful immersion of the actor in the character would offer the audience an emotionally transcendent experience. The Seoul production of *Please Turn Out the Lights* was clearly based on such principles. In fact, in terms both of dramaturgy and of production, Yi's play is almost entirely guided by Western notions, and none of its elements can be traced back to the heritage of traditional Korean theatre or performance arts.

But for all of these formal aspects of *Please Turn Out the Lights* that seem wholly Western in conception and execution, Yi's play is nonetheless decidedly Korean in character. To see it as a case of a foreign hegemonic model imposing itself on those who produced the play and imposed by them in turn on their audience is to misunderstand fundamentally the dynamics of cross-cultural phenomena. Cultural performances that encompass nonindigenous forms and content are always marked by the interaction of the familiar and the alien, even if the familiar is brought into play only by the expectations and mind-sets of the indigenous audience.

Poesis, as Michel de Certeau has shown in his landmark book *The Practice of Everyday Life,* involves first of all the manipulation of a text by users who are not its makers.[10] Setting aside the common assumption that consumers are passive and guided by established rules, de Certeau argues that consumption is instead "another" production and characterizes it as "devious, disperse but insulated everywhere silently and almost invisibly" because consumption "does not manifest itself through its own products, but rather through its ways of using" (xi-xii). He claims that such user tactics "lend a political dimension to everyday practices" (xvii). De Certeau's insistence that consumption is active and that consumers of texts—and plays—are also producers offers a useful way of thinking about Korean response to Yi's Westernized play. If de Certeau is right, then all those spectators are not just sitting there inertly absorbing what they see and hear on stage, but rather are energetically processing it, making sense of it, and more importantly, converting it to their own uses. From this perspective, *Please Turn Out the Lights,* however Western its appearances, is in its enactment for Korean audiences and in its use of the Korean language a theatrical event at least as Korean as it is Western. Although this drama is clearly based on a Western model, we cannot assume it uniformly instills "Westernness" in the indigenous audience. In other words, the presence of Western dramatic elements does not indicate, much less prove, that the Korean ways of thinking or Korean culture has been squashed.

There are many aspects of the play and its production that demonstrate how Korean readings can emerge even at the most seemingly Western moments. Particularly striking is the way the play handles its female characters, who, despite a Western gloss, remain indisputably wedded to Korean gender roles. This play is concerned with the theme of redemption, with finding a way of escaping the past. Yet, though both male and female protagonists have suffered equally from past events partially beyond their control, here the object of redemption is only the male character. On the one hand, Kang's resignation from public office is prompted by his guilt. On the other, it heralds his emancipation from the dominance of social institutions that care little for individual happiness. His decision to visit his long-lost lover on the crucial night shows that his desire for liberation has finally caused him to act. His resignation, visit, and confessions free Kang from an emotional burden and thus offer him an important opportunity to renew himself. But for Pak no such moment of redemption occurs. She neither takes such an initiative, nor does she give evidence of any significant change within herself. In fact, she only remains what she has always been, a victim, now passively accepting what Kang thrusts upon her, which is not so much a new life as a new role, that of savior, as one capable, through the sexual ritual implied in the play's closing lines, of bringing about forgiveness and offering the promise of a new life for Kang. Although Kang's sins are resolved through this meeting, nothing in her reunion with Kang addresses the violence against Pak, raped by her now estranged husband. Whatever happiness she herself might subsequently find is predominantly determined by the male protagonist's transformation, for once again the female protagonist is not given an opportunity to redeem the wounds of the past or reassert her self-esteem. In the last scene the male protagonist's internal changes are revealed:

KANG. Each time my brother calls me daddy, it feels like an eternal punishment; one that I'll never be free of. If the truth is disclosed, the world will look down upon my mother. I'll feel ashamed too. But, then I should be able to raise my face to have a fresh look at my mother, for there's no need to disguise any more. The truth is . . . it's her love that made me come this far. She endured shame and condemnation to raise us.
PAK. It seems you took such a long detour. It could've been easier, you know.
KANG. It's time for you to carve out the old wounds with purifying energy.
PAK. Purifying energy?
KANG. Every creature on earth has this energy. It has a self-purifying effect. Even in the circumstances of social, institutional distortion, this energy always directs us back to our essence. It's an instinct. No other energy can override this one.
PAK. Over the years I've been telling myself it's best to forget the person who doesn't belong to me. It's like I appreciate flowers in the fields; they don't have to belong to me. Just being there . . . they're . . .

A sound of music, of a soft and tranquil tone, is heard in the background.

KANG. We need to repot.
PAK. Repot?
KANG. I'm trying to rekindle the flame, as it says in your poem.
PAK. I see. A history can begin anew even on the field of burned grass.
KANG. I've got so many words to say to you, but they're circling within my heart. I can't express my thoughts.
PAK. You don't have to speak. It is understood in my heart.
KANG. We all have a destination.
PAK. As if there's truth in all of us?
KANG. Yes. Whether you want to be a poet, a nun, or an artist.
PAK. Please turn out the light.

The lights fade out. (75–76)[11]

One might claim, with some justification, that Pak here fills the Western role of lover-savior; she becomes a Gretchen-like embodiment of Goethe's "eternally feminine," which, when combined with male striving, can effect redemption. To place *Please Turn Out the Lights* once again in the tradition of Ibsen's drama, Pak is also a Korean Solveig who "eternally" waits for the return of her lover, Peer Gynt, who finally comes back to affirm the virtue of human relationships. But Pak's role is determined by some fundamental axioms of traditional Korean culture, in which the needs of women are nearly, if not entirely, subordinated to those of men. Within the play, Pak achieves Kang's transformation largely by her silence; though just as troubled by the past as Kang, she never fully reveals her ordeal, and is thus never in a position to experience redemption through such a revelation. In light of the fact that her present crisis was largely caused by Kang, and that she is as much a victim as he is, she not only needs, but is entitled to, salvation. However, Pak remains silent, leaving her own account of the rape and abuse unspoken. She becomes a man's savior by abandoning her own needs, desires, and even identity. Thus she fulfills the demands made on women by the massively patriarchal culture of Korea. And herein lies one of the play's great paradoxes, for while Kang seeks the freedom of individuality that a culture advocating traditional rather than personal values has denied him, he does so expecting his female counterpart to continue to play a role defined by tradition. She must perform a redemptive ritual for him, but he need not do so for her. Thus for all of its apparent modernity, *Please Turn Out the Lights* also advocates specifically Korean traditional patriarchal practices and values. Pak's identity is precisely determined by countless other women in Korean history and literature; if she has a Western identity at all, it is an utterly superficial one. A very complimentary review of the play by Hô In-hwa that appeared on January 10, 1992, in *Kyông-hyang sinmun* [*Kyông-hyang Newspaper*] concludes:

> For one hundred and fifty minutes the audience is awed by the unconventional usage of Korean and participates in director Kang Yông-gôl's project of reshaping the Korean sensibility. The protagonists, who were bound by the institutions to suppress their instincts, are reassured of their love that had started ten years ago. The female protagonist's final utterance, "please turn out the lights," signals the beginning of a new life, a promising future, and implies what can be "turned on" by "turning out."

But despite the review's optimistic assessment of the impact of the Western-inspired treatment of Pak, Korean audiences had ample scope to read Pak in light of their accepted values and norms.

Such considerations draw our attention to the treatment of another female character in this play, Kang's wife, who appears on stage and is frequently referred to. What is ironic about her is first of all that she has no name; in the text she is simply "the Wife." While her existence may be important to the play's development, her personality is indeterminate, perhaps even suppressed. Myông In-sô, a drama critic and reviewer, rightly points out that except for Kang and Pak, "other characters are not sufficiently portrayed, especially Kang's wife and Pak's husband. Their characterization is flat, and their function is one dimensional. They are merely instruments to make the protagonists come alive."[12] As mentioned earlier, Kang's wife would have been expected to make a "good match," unlike Kang, a menial construction worker whose education and family background were immensely inferior to hers, and who had in addition a small child to support. The woman nonetheless developed compassion and even fond feelings for Kang, and this led to their marriage. Kang recalls his first meeting with his future wife as follows:

> KANG. I met her at the construction site.
> PAK. You mean, your wife?
> KANG. It was one evening. A fancy car drove in, and the president and his daughter got out. I was throwing up blood in the corner. The daughter saw me. From the next day on, she came to visit me every day.
> PAK. It must have been fate.
> KANG. Isn't it strange?
> PAK. What?
> KANG. My wife. A marriage with her was out of the question. I did my very best to brush her off. I even exaggerated my past, but she didn't give up. (72)

In this scene the audience learns about the peculiarity of Kang's first meeting with his wife. But what is truly peculiar about their relationship is that even now, ten years into the marriage, Kang is still baffled by his wife's willingness and sacrifice. Kang feels that the shower of generosity and kindness alienates him from his true identity. From their first meeting, Kang's wife served as a shelter to Kang and his "son." Aided by this extraordinary circumstance, Kang was able to escape his misfortunes and achieve material success.

The irony is heavy: Kang's wife is made obsolete by the very ideology that constitutes her womanhood. In light of the unfortunate turn of events waiting for her, the point seems to be that in an ever-expanding, ever-changing Korea, the traditional values founded on the long-standing Confucian doctrine fostering the ideal of a good mother and caring wife are no longer honored or considered binding. With the advent of modernization, a transformation in traditional patterns of marital and interpersonal relationships has taken place. Kang's wife, who is the embodiment of a definitive cultural value, is to be replaced by Kang's mistress. Modernity demands a new breed of woman, one who can rehabilitate the failing male protagonist so that he can experience emotional gratification. As we have seen, the emphasis in this play is on the central character's transformation, and the new role of the woman is to help him achieve it.

It is clear that in the course of massive cultural changes, femininity is one of the first values to be placed under imminent reconstruction. However, as they are reflected in this drama, the actual historical efforts to redefine femininity scarcely go beyond the "replacement" or "exchange" of women that, in Lévi-Strauss's account, takes place in tribal society for economic and reproductive purposes. The "barter" occurring in *Please Turn Out the Lights* serves to reinforce the dominant patriarchal rule. Pak, although no longer required to enact a robotic femininity, as was Kang's wife, is expected to fulfill the requirements of a new kind of femininity that, while different in many characteristic details, is yet defined by a value system in which women's role is ultimately enunciated from a patriarchal position.

Please Turn Out the Lights is a text calling out for a feminist critique, and for many Western readers such a criticism would be quite acceptable, if not expected. From the vantage point of feminist criticism, Pak and Kang's wife, instead of submitting to the needs of the male order, should, like Ibsen's Nora, have proclaimed their freedom at the crucial moment. The wife's blind devotion and the mistress's forgiveness are not what Western audiences would normally expect to experience. A Nora-like response by one or both of the women would indeed make it a feminist play but would also, so it would seem, make it out of place in the modern Korean theatre. Many members of the audience—men and women, students and workers—appreciated *Please Turn Out the Lights* for its sentimentality and emotions. It does not appear that they understood where the play was coming from, or assumed a critical stance vis-à-vis the play's ideological bases. This point perhaps accounts for the play's great success—particularly among the men who flocked to it.

The fact that no feminist critique of the social and ideological realities shown in the play emerges, either within the play itself or from audience or reviewer responses to the play, suggests that texts as apparently Western as Yi's play can serve quite local and non-Western functions. Seen from this perspective, critiques of such texts inspired by a postcolonialist theory and practice, however well intended and however perceptive in many ways, ultimately miss the "forest" of the pervasiveness of local culture by concentrating on the "trees" of Western influence that are planted in it. Korean nationalist critics have also been unable to see how a play like Yi's, commercial in its orientation and Western in its production, can nonetheless still be seen as a strong manifestation of a local tradition. The majority of Korean nationalists are elitists who advocate highbrow art. They distinguish between literary (highbrow), commercial (middlebrow and bad), and popular (lowbrow) art. A typical elitist criticism is that to the extent that it is a product of Westernization, a drama like Yi's cannot express anything about the indigenous culture and population but instead represents the loss of Korean cultural heritage. More important, nationalist critics claim that "popular" culture, because of its lack of social and historical consciousness and because of its susceptibility to Western influence, never challenges the political status quo and thus, being unable to offer a direction for progress, can only subscribe to existing norms.

It is certainly true that *Please Turn Out the Lights* perpetuates certain oppressive hegemonic practices, but those practices are largely Korean, not Western. This point emerges clearly when we consider the composition of the audiences for this smash dramatic hit. As I have already suggested, *Please Turn Out the Lights* has attracted a much wider group than the largely young and female audiences that attend most plays of this sort in Seoul. Male, white-collar workers, especially young and middle-aged ones, have attended the play in large numbers. This suggests that in some way the men in attendance are identifying with Kang; indeed there is good reason for them to do so, since in many ways he embodies the experience of their generation. Like Kang, male members of this generation have felt themselves torn between demands that they accept traditional ways of life, ways that deny importance to individuality and personal happiness, and the desire for a sense of personal satisfaction. To them the play offers the only imaginable kind of solution, one that is traditionally Korean, affording these men a "redemption" acquired at the expense of denying individuality and liberation to women. This solution entails a reinforcement of the traditional role of women in Korean culture. The men in the audience may not even be aware of this, but there is no evidence that the traditional female audience took offense, either. The matter never surfaced in reviews or in public response. What, on the surface, appeared to speak to new social situations in a highly Westernized style ultimately proved to be nothing but the voice of the past reasserting once again the traditional and the patriarchal, only this time in a guise that on first consideration may seem exotic, alien, imported, and Western.

In the end, then, we can see the working of a powerful traditional cultural discourse even in a play as modern and Western as *Please Turn Out the*

Lights. The female protagonist in the play may look modern, but more prominent than any question of her freedom is the theme of her sacrifice. Emancipation in the true sense is not possible for her because she inhabits a world that is both morally repressive and politically conservative. Pak finds herself a new role, but one that ultimately affirms the dominant status of the male order. In light of this characteristic, *Please Turn Out the Lights* is at once a Korean text and Western as well.

Western critics often view the spread of Western capitalism as a form of cultural imperialism. The question of the relation between economic and cultural domination is a sensitive issue, but the general view is that transnational corporations, largely based in the United States, which are developing global investment and marketing strategies, constitute the principal agent of cultural homogenization. Such critics believe that cultural synchronization is a threat to the cultural autonomy of the non-Western and that the ready acceptance of Western models implies the simultaneous imposition of Western values and ideology on indigenous people. Many Korean critics have likewise argued that the existence of Western imports, especially in a politically marginalized nation such as Korea, jeopardizes the preservation of the cultural heritage and native worldview. But, as I have tried to show, both Western and Korean critics have underestimated the resilience of indigenous cultures.

It is true that twentieth-century Korean theatre is thoroughly modeled on the realist theatre of the West. For this apparent similarity, the Korean theatre has been accused of lacking political consciousness and thus posing a threat to traditional culture. As the example of *Please Turn Out the Lights* shows, however, Korean audiences can interpret and appreciate a play, however Western, according to their own traditional morals and worldview.

A play like *Please Turn Out the Lights* asks us to rethink commonly held assumptions about East-West relations. Just as we, as individuals in the audience of drama and culture, are all members of various interpretative communities, so no culture represents a single nationhood or cultural identity. Culture constantly undergoes changes, tending toward fluidity rather than stasis. The resemblances between Korean plays and their Western counterparts occur no doubt partly as a consequence of the global spread of capitalism and the frequent exchange among nations. Such plays imply a fusion of cultures, in which differences and similarities are negotiated or eventually amalgamated. I view this process as positive. The sense of assertiveness of Korean values found in *Please Turn Out the Lights* challenges the discourse of cultural imperialism, mapping out a space from which we can begin to imagine, theorize, and articulate the indigenous culture as an agent able to act in its own interests.

Notes

1. See, for example, Fredric Jameson, "Third-World Literature in the Era of Multinational Capitalism," *Social Text* 15,3 (1986): 65–88, and "Postmodernism, or the Cultural Logic of late Capitalism," *New Left Review* 46 (1984): 53–92, as well as Anne McIntock, *Imperial Leather* (New York: Routledge, 1995).

2. See Chungmoo Choi, "The Discourse of Decolonization and Popular Memory: South Korea," *Positions* 1 (1993): 77–102, and Nak-chung Paik, "The Idea of Korean National Literature Then and Now," *Positions* 1 (1993): 553–580.

3. Admission to two other plays was W8,000 ($10), to four more W7,000 ($8.75). One play was billed at the lowest price, W6,000 ($7.50), and the highest was *Agnes of God* at W15,000 ($18.75).

4. Yông-gôl Kang, "Muôti changgi kong'yônûl mandûnûnga?" ("What Makes a Play Successful?"), *The Korean Theatre Review* 196 (1992): 12–15.

5. The play was staged twice a day from Thursday through Sunday; a matinee at four o'clock and the evening show at half past seven. Tuesdays and Wednesdays there was an evening show only; there were no performances on Mondays.

6. Man-hûi Yi, *Pul chom kkô chuseyo* [*Please Turn Out the Lights*] (Seoul: Taehakro Kûkchang, 1992), 13.

7. Yun-ch'ôl Kim, "*Please Turn Out the Lights*: The Strengths, Weaknesses, and Flaws of Experiment," *The Korean Theatre Review* 194 (1992): 13.

8. It is not clear when and how T'alch'um became a popular form of entertainment, but it is thought to have grown out of agricultural rituals. T'alch'um used to be performed on major Buddhist holidays and at planting and harvesting times. Sarcastic in nature and poking fun at the landed gentry, T'alch'um was performed by amateurs—usually farmers. Dance is central to the performance, and dialogue supplementary. T'alch'um tells its story through stock characters of perverted monk, villainous gentleman, mysterious shaman, traveling minstrels, sharp-tongued servant, and virgin. Nong'ak is entirely a dance piece, in which the dancers twirl twelve-feet-long, thin, white tape attached to their hats to the fast beat of drums, gongs, and conical flutes. Nong'ak is believed to have been performed as part of shamanistic rituals to protect the villagers from natural and human disaster and to give thanks to the spirits of the surrounding mountains and rivers. Today both T'alch'um and Nong'ak are by and large treated as types of spectator entertainment, even if they are performed outdoors. In the past, spectators were encouraged to join in the dance to become part of the performance.

9. Since attending a performance of *Please Turn Out the Lights*, I have seen several photographs of this scene, in *The Korean Theatre Review* and in major newspapers. It seems that the props were changed from time to time, if not for each performance.

10. Michel de Certeau, *The Practice of Everyday Life,* trans. Steve Rendall (Berkeley and Los Angeles: University of California Press, 1988), xii–xiii.

11. All the translations of *Please Turn Out the Lights* in this essay are mine.

12. In-sô Myông, "Taejungkûk ûro chari chapûl su ittnûn mudae" ("The State that Can Claim the Mainstream"), *The Korean Theatre Review* 206 (1993): 35.

Chapter 6

Representing the Chinese
Nation-State in Filmic Discourse[1]

Sheldon H. Lu

M edia theorists and cultural critics have argued that the post–Cold War era is the age of transnational media and cultural globaliza-tion. Transnationalization, in this formulation, breaks down na-tional barriers and penetrates the remote corners of the globe. Globalization, as succinctly defined by Roland Robertson, "refers both to the compression of the world and the intensification of consciousness of the world as a whole."[2] This space-time compression has in part been brought about by the spread of new communication technologies across the globe. Residents of Third World countries can gain easy access to the cultural products of the First World such as film, TV programs, popular music, and fashion. At first glance, it would seem that the primacy of the nation-state is fading and that cultural production and consumption occur frequently at the local, transnational, and global levels.[3] Such a view seems to be confirmed by the global box-office success of the Hollywood block-buster *Titanic* in 1998. This film has been widely seen all over the world, including in Chinese-speaking communities—mainland China, Taiwan, and Hong Kong—and has even won the praise of Chinese president Jiang Zemin.[4]

As *Titanic* suggests, the transnational distribution of Western cultural products appears to have the power to transcend geographic barriers and ideological differences. For example, the Disney animation film *Mulan* (which had been banned by Chinese censors because of Disney's produc-tion of *Kundun*, which touched on the sensitive topic of Tibet) was finally released in mainland China in 1999, one year after it had been screened elsewhere in the world. As a contemporary Western high-tech rendition of a centuries-old Chinese folktale, the film has now entered the market of the original home country, seemingly offering further proof of the invincibility of transnational media. *Mulan,* many might argue, reveals how the memory and deep history of indigenous cultures are being flattened out and appropriated by giant media corporations, with the result that subjectivities and cultural identities across the globe are reconstituted in accordance with the operations of capital and the market. In the process, many might also claim, identity formation has increasingly become a "horizontal" process—in which identities are traded like commodities across national borders—as opposed to being a "vertical" construction based on the time-honored traditions of local communities and their status hierarchies.

Yet theorists also point out that the irreversible process of modernization/ postmodernization does not necessarily mean the eclipse of the nation as a key feature of cultural formations. Mike Featherstone, for one, argues that "theories of modernity which emphasize a relentless process of instrumental rationalization which effectively 'empties out' a society's repository of cultural traditions and meanings are misconceived."[5] In this essay, I want to expand on such claims, showing, in particular, how mainland China and the West have drawn—not always in a positive way—upon history, memory, and tradition in the imaging and imagining of modern China as an object of representation in film. As I intend to demonstrate, the portrayal of China and of Chinese identity in both Hollywood and mainland films has not been able to escape from the trap of national and political stereotyping. In both sets of films, in fact, the East does not meet the West in any mutually enriching way—that is, in a way that would move beyond clichéd national images. Nor is identity understood to be a fluid construction not dependent on national ties. Only in films made in Hong Kong during the period under discussion, as I shall mention, is the possibility of a truly transnational perspective explored.

The historic moment of the return of Hong Kong to China in 1997 aroused tremendous worldwide interest in questions about China and Chinese national identity. Who is Chinese? Who is Hong Kongese? What does China or "Chineseness" refer to? In a number of films and other popular cultural products made in 1997, China as a nation-state and a geopolitical entity and Chinese/Hong Kongese as a national and cultural identity became the subjects of representation. In this study, I briefly examine three clusters of films released since 1997: Hollywood films about China and Tibet; mainland Chinese films about modern Chinese history; and Hong Kong films about Chinese/Hong Kongese identity. My purpose is to discover and compare the critical assumptions within these films about nationality and identity in regard to the subject of China. This critique is especially important given that, due to censorship, political control, and film distribution networks, Chinese and American audiences do not get to watch the films of the other countries, and thus remain for the most part blind to other points of view.

Hollywood's China

In 1997–98, a series of highly publicized Hollywood films related to China were released: *Kundun, Seven Years in Tibet, Red Corner,* and *Chinese Box.* All these films depict various ethnicities and employ an international cast of actors and actresses; in that sense, they seem to be transnational. Yet all four are premised on the notions of the nation-state, fixed territoriality, and sovereignty. In fact, a residual Cold War politics informs these Hollywood productions, as is obvious from their consistently heavy-handed recourse to anti-Communist sentiments or cultural stereotypes.

Film critics based in the United States have not failed to notice the overtly political intentions of *Red Corner.* As a matter of fact, the film and its lead actor, Richard Gere, have received some of the worst and most disparaging reviews for a big-budget Hollywood film. Roger Ebert flatly states that "*Red Corner* plays like a xenophobic travelogue crossed with Perry Mason." "To some degree," Ebert continues, "Gere set himself up by appearing in this film; as an outspoken critic of China and follower of the Dalai Lama, he has a case to plead. It's surprising, then, that he chooses to do it so lamely in such a lugubrious movie."[6] In his review article "*Corner:* A Heavy-Handed Battle with Justice in China," Kenneth Turan characterizes the film in a similar way: "Its one-man-against-the-system story is hackneyed and the points it thinks it's making about the state of justice in China are hampered by an attitude that verges on the xenophobic." For Turan, the Chinese as presented in the film "are among the most sinister and unsmiling group of Asians to emerge from Hollywood since the 'Beasts from the East' movies of World War II."[7] In her review "*Red Corner:* Melodrama-cum-Credibility Snag," Janet Maslin points out: "*Red Corner* shows an earnest, committed interest in criticizing Chinese totalitarianism (the film's opening date, timed to President Jiang Zemin's visit, is apparently no accident) even as it piles on some of the standard twists and turns associated with cinematic trial stories."[8] As these reviewers' comments suggest, the heavy dose of anti-Communist China ideology and the forceful performance of Richard Gere diminish the film's credibility.

Kundun, a visually stunning work released in the same period, is nevertheless troubled by similar problems. It is perhaps appropriate that *Kundun* is directed by none other than Martin Scorsese, a former altar boy in a Roman Catholic church, who once contemplated becoming a priest, and who made the film *The Last Temptation of Christ*. The lofty, spiritual aspirations of *Kundun* are what make the film beautiful and compelling. However, as critics have noted, this deep spirituality "denies the Dalai Lama humanity; he is permitted certain little human touches, but is essentially an icon, not a man." The enchanting visuals and music are "an aid to worship: the [film] wants to enhance, not question."[9] Religious iconography thus takes over and any chance for a realistic representation is lost.

The simplistic portrayal of pious Tibetans and the gross caricature of the Chinese in the film also stand in the way of an engaging drama. One reviewer states that "in the film's most jarring sequence, the boy, now a teenager, travels to Beijing for a meeting with Mao Zedong, whom Robert Lin plays as a shrill, campy caricature." For him, "Scorsese has made a film that is as much a prayer as it is a movie."[10] In the words of another reviewer, "Both *Kundun* and the earlier and sillier *Seven Years in Tibet* have had trouble making Tibet's plight as moving as it ought to be. And *Kundun*'s decision to turn Chairman Mao (Robert Lin) into an oddly fey and borderline campy maximum leader does not help the situation."[11]

The viewer cannot help but compare *Kundun,* a biopic of the fourteenth Dalai Lama, with the highly successful epic film *The Last Emperor* by Bertolucci. The lives of these two historical figures are similar in many respects. What makes *The Last Emperor* a much more credible and moving tale is precisely the depiction of the emperor from early childhood to old age as a deeply human figure caught in complex historical events, as well as the avoidance of stereotypical representations of Chinese characters, whether they are communist officials or loyal subjects of the emperor. Unfortunately, *Kundun* fails to achieve a similar complexity, perhaps in part because of its allegiance to nationalist stereotypes and to an overt political agenda.

Hollywood's fascination with Tibet/China in this period resulted in another major picture: *Seven Years in Tibet*. The film features the adventures of Heinrich Harrer, an Austrian mountain climber in the 1930s and 1940s. Like *Red Corner,* which showcases Gere, the film is essentially a star vehicle for Brad Pitt, who plays the role of Harrer.[12] The story details Harrer's departure from Austria for Tibet, his mountain climbing in Tibet, the hardships he endures as a prisoner in a British concentration camp during World War II, and finally his return to Tibet and his meeting with the Dalai Lama after the war. Despite its focus on Pitt/Harrer, the film does not escape the same anti-Communist China passion that enfeebles *Kundun*. Toward the end of the film, the director inserts horrendous scenes of the invasion of Tibet by masses of Chinese soldiers at a time when Harrer had already gone back to Europe after his "seven years in Tibet"; the invasion scene has, in fact, nothing whatsoever to do with this period of Tibetan history. In the words of one film critic, what is lamentable is that "perhaps suspecting that his audience is falling asleep, [director Jean-Jacques] Annaud ends *Tibet* with violent but pointless scenes of the 1950 Chinese invasion."[13]

Chinese Box, directed by Wayne Wang, which tackles the subject of Hong Kong's reversion to China, also suffers from heavy-handed political allegory and a simplistic delineation of East-West cultural dynamics. The film depicts the lives of several people living in Hong Kong on the eve of its return to mainland China. John (Jeremy Irons), a British journalist who will die of leukemia in a matter of six months; Vivian (Gong Li), a night-club hostess and an ex-prostitute from mainland China; Jean (Maggie Cheung), a Hong Kong street girl with a scarred face—these characters are, respectively, none too subtle symbols of Great Britain, mainland China, and Hong Kong. Early in a party scene in the film, a student-democracy fighter shoots himself to death to warn the residents of Hong Kong of the impending loss of their freedom that is assumed will occur after the handover of Hong Kong to China. Clearly the film is siding here with a pro-Western, anti-China position in a way that falls back on stereotypes and unexamined assumptions about East and West.

In addition to its overt political symbolism, the film also harks back to Orientalist representations of Hong Kong and the East. Reviewer Kevin Thomas reminds the viewer: "John is the dashing, reflective but impassioned Anglo-Saxon and Li's Vivian the dazzling Asian beauty from countless movies and plays past. Anyone hear strains of *Love is a Many-Splendored Thing*? Echoes of *Suzie Wong*?"[14] Elements of Cold War ideology, East-West political antagonism, and entrenched Orientalist habits of thought indelibly inform this Hollywood film.[15]

Indeed, it seems that politicizing cultural productions from China has become a habitual way of dealing with postsocialist China in the West, even in the post–Cold War era. See, for instance, a simple advertisement page in the *New York Times*. In May 1999, there appeared an advertisement for a new film about China, *Xiu Xiu: The Sent Down Girl*, directed by the former Chinese actress, now a Chinese-American actress, Joan Chen (Chen Chong). The advertisement includes the words: "BANNED IN CHINA FOR SEXUAL AND POLITICAL CONTENT."[16] As this sentence suggests, the worth of this film seems to depend on its being banned in China. Not surprisingly, the advertisement adopts the marketing strategies of politicization and eroticization in order to arouse the public's interest in a film about Red China.

Joan Chen provides a particularly instructive example of the apparent possibilities but also very real limits of cross-cultural flow and horizontal identity exchange. The actress has appeared in both Chinese and Hollywood films and is well known in both countries. This time, in *Xiu Xiu,* she turns into a director. This change in role is applauded by Richard Corliss, an American critic, in his review of the film. Yet his words tellingly reveal an extremely limited horizon for filmmaking about China in the West. His remarks are noteworthy, and indeed psychoanalytically uncanny:

> Clearly, Chen's striking beauty—searching eyes, long, strong neck and, it must be said, the most luscious mouth on either side of the Pacific—is merely the wrapping for surpassing talent and drive. Hollywood's favorite China doll wanted to direct.[17]

On the left side of the page featuring the article is a full-page frontal picture of a smiling Joan Chen, with the caption "Joan of art." In making this new career move from symbol of Asian femininity to director, Chen cannot escape Western (male) assumptions that effectively typecast her as "China doll," even as she moves behind the camera into the director's chair. Once again, film production about China must conform to a familiar formula of politics plus sex plus exoticization, the recipe needed in order to be successful in the West.

This manner of wishful thinking or willful misunderstanding is well explained in the words of Orville Schell, in his review of *Windhorse*, a low-budget American film partly shot in Tibet by Paul Wagner. Schell writes the following about the Western fantasy of Tibet:

> We tend to see Tibet through one of the most powerful utopian mythologies of this century—of a fabled, Shangri-La-like refuge from the outside world. . . . So deeply rooted in the Western psyche is our version of Tibet that it has become virtually part of our mental DNA. One way or another every article, book or film on Tibet must contend with the political and cultural determinism of this myth. . . . Although old Tibet has all but vanished and Lhasa is turning into a charmless Chinese city, we Westerners still love to retreat into virtual versions reconstructed on film sets.[18]

Be it Tibet or China, preexisting modes of perception still play a large role in the politics of cultural production and exhibition in the post–Cold War era. To what extent do such perceptions reveal more about the self rather than the other? Surely it is not difficult to see that the Hollywood films that I briefly discussed above are as much expressions of the American frame of mind as they are representations of the reality of a foreign country. It may be the case that as long as the nature of the Chinese state remains the same, or as long as the memory of the 1989 Tiananmen incident does not fade, the Western public's attitude toward China—including what it wants to see and know about China—will not alter significantly. Indeed, it is no exaggeration to say that China has become an integral part of

American domestic politics as the country looms large as a potential competitor to the United States in terms of geopolitics, global trade, military buildup, and international influence at the turn of the millennium. The "China question" has become a staple campaign topic that must be passionately addressed by every presidential candidate. Certainly the demonization of China is a cheap and easy political move calculated to show patriotism and loyalty to America and to fortify the moral supremacy of Western culture. The media, the Congress, and politicians have made sure that the ordinary American citizen is inundated with endless stories of Chinese espionage, illegal presidential campaign donations, human rights violations, and so forth. In the realm of popular culture, Hollywood films like *Kundun, Red Corner, Chinese Box,* and *Seven Years in Tibet* play a part in the perpetuation of an image of China as the demonic other. In such films, we seem very far indeed from any representation of transnational culture or identity.

Films from Mainland China

To celebrate the return of Hong Kong to the motherland, China produced a flood of artistic works in various forms and media: drama, music, TV programs, and films. Most noticeable are two big-budget epic films—*The Opium War (Yapian zhanzheng),* directed by Xie Jin, and *Red River Valley (Hong he gu),* directed by Feng Xiaogang. Both films attempt to chronicle the military interference in China as a result of British imperialism and to narrate the painful and difficult emergence of China as a would-be modern nation-state. *Red River Valley* further involves the triangular relationship among China, the West, and Tibet, and describes the plight of China as a multiethnic nation. As we might expect, the Chinese films' treatment of these issues is very different from the American films mentioned earlier. Nevertheless, although these films are diametrically opposed to those American films in ideology and historical representation, they all are premised on the notions of the nation-state, fixed territoriality, and sovereignty.

At both the beginning and the end of *Red River Valley,* the voice-over, the voice of an old (Tibetan) lady, tells an ancient legend: Goddess Mount Everest gives birth to three brothers who are best friends: the eldest is the Yellow River, the second is the Yangtze River, and the youngest is the Yarlung Zangbo (Yalu Tsangpo) River. In such a manner, the film establishes a relationship between the Chinese (Han) and Tibetans as grounded in nature and blood. The narrative hinges on a romantic love story between Tibetans and Chinese. In the beginning of the twentieth century, there is a severe drought along the Yellow River. A Chinese (Han) daughter named

Xue'er is to be sacrificed and thrown into the river during a prayer for rain. She jumps into the river and manages to escape, and is later rescued by an old Tibetan woman. She then lives among the Tibetans, takes the name Xue'er Dawa, and falls in love with a Tibetan youth, Gesang. In such a manner, the film as a whole repeatedly confirms that Hans and Tibetans in fact belong to one Chinese family and are mutually indebted to each other.

In contrast, the British expedition troops and Colonel Rockman are described as the evil force that slaughters the Tibetans. The British attempt to break up the great Chinese family of nationalities and take Tibet away, but they fail. The only British person who still has a conscience is the journalist Jones. At the end of the film, he says, "Why should we change their world with our world? This is a people that can never be conquered and destroyed. Behind its back there is an even vaster land; that is the East that we can never conquer." Through the words of a Westerner, the film affirms the independence and solidarity of the peoples of the East. But at the same time, it reinforces Eastern stereotypes about a demonic West and constructs a barrier between the two cultures.

A noteworthy event was a transnational, East-West romance during and after the shooting of the film. Ning Jing, the mainland actress playing the part of the proud daughter of a Tibetan leader, and Paul Kersey,[19] the American actor performing the role of Jones, fell in love, married each other, and gave birth to a child. Ironically, this love story outside the text of the film deconstructs the East-West ideological antagonism within the filmic discourse. In this instance, the film's xenophobia and allegiance to the ideal of the nation-state are strikingly at odds with what was happening in the actual lives of at least two people. The courtship of Ning Jing and Paul Kersey points to a discrepancy between reality and representation, between the bonding of people across national boundaries on the one hand and a monologic cinematic discourse in the name of the nation-state on the other. (Subsequently, the husband-wife team starred in a 1999 film about an East-West, cross-cultural romance set in China during World War II, *Romance of the Yellow River* [*Huanghe juelian*], which was the official entry submitted by China to the Motion Picture Academy of Arts and Sciences to compete for the Oscar for best foreign film.)

At the time of Hong Kong's handover, another widely publicized film was *The Opium War*, which narrates the impotence of the Qing court, the encroachment of China by the West, and the loss of Hong Kong to Britain. In 1959, the mainland-produced film *Commissioner Lin* (*Lin Zexu*) depicted the same set of historical events. The last part of the 1959 film described the spontaneous uprising of the peasants in Sanyuanli against foreign invaders, an event that heralded a century of revolution in China culminating in the Communist revolution. In the new film directed by Xie

Jin, this grand narrative of the people's struggle against imperialism is absent. The film adopts a different narrative strategy. It probes into the state of isolation and ignorance of the Qing dynasty vis-à-vis the West at the dawning of the modern age. Despite the steely determination and bravery of the Qing people to end the opium trade—from the emperor on down through the ablest ministers (Lin Zexu included), the generals, local officials, and even ordinary citizens—China cannot avert a losing battle against a fully industrialized modern nation, Great Britain. On a par with the policy of "reform and openness" of the current regime, the film implicitly calls for modernization and globalization so that China may become a strong, independent country among the nations of the world.[20] Despite this laudable agenda, *The Opium War* cannot escape from clichéd images of the "battle of nations" or from the strong nationalist sentiments that undergird such images.

A precursor on the subject of Tibet was the 1963 mainland film classic *Serfs (Nongnu)*. The film portrays the Han people of China as the liberators of the Tibetans, and the People's Liberation Army as the savior that frees Tibet from feudalism and slavery. Han/Chinese society is supposed to be at a higher stage of civilization, whereas the Tibetans are seen as living at a lower, primitive, barbaric stage of human history. Again, *Serfs* narrates the grand tales of liberation and class struggle. *Red River Valley*, in contrast, does not criticize the old social system, nor does it make an attempt to unveil the plot of the historical evolution of humanity from a Marxist, Maoist, and revolutionary perspective. Instead, it painstakingly depicts a natural, deep, intimate relationship between the Chinese and Tibetans that transcends class, social, and institutional barriers. It is an ahistorical affirmation of the unity of the Chinese family of nationalities.

Mainland films such as *Red River Valley* and *The Opium War* stress the integrity and continuity of China as a multiethnic nation and condemn the invasion of the Chinese sovereign nation-state by the colonialists. In contrast, Hollywood productions such as *Kundun, Seven Years in Tibet,* and *Red Corner* advocate the moral superiority of the West and Tibet over the Chinese, and castigate the Chinese rule of Tibet. The two sets of films cannot be further apart ideologically. Nevertheless, they share some basic assumptions—the integrity of the nation-state, fixed territoriality, the inviolability of sovereignty, and the idea that cultural and political identity derives from identification with the nation-state. In these films, a Cold War-style ideological opposition between the East and the West seems rooted in a shared model of geopolitics as well as in a common conception of fixed, unchanging national borders. Judging by these films, our present historical moment, the so-called era of transnationalism, falls far short of the utopian vision of a "borderless world."

Films from Hong Kong

Films made by Hong Kongese themselves around the handover period offer a different take on Hong Kongese and Chinese identity and, more importantly, manage to escape from the bind in which both Hollywood and Chinese films find themselves. *Comrades, Almost a Love Story* (*Tian mimi*), by Peter Chan, and *Happy Together* (*Chunguang zhaxie*), by Wong Kar-wai, are particularly striking examples.[21] *Comrades* tells the story of two mainland immigrants in Hong Kong, a man and a woman. The film details the events of their work, life, love, and their eventual reunion in the United States. *Happy Together* relates the life, travel, and love of two gay Hong Kong men in South America. The two films do not rehearse the grand historical narratives about the formation of the Chinese nation-state, but unfold the little tales of ordinary citizens. Here, Hong Kongese/Chinese identity becomes a mobile, deterritorialized, transnational, and changing phenomenon. In both films, Hong Kongese/Chinese identity does not depend on one's place of birth or residence: one does not have to live in China or Hong Kong, be an active member of a nation-state, or define oneself in political, territorial, and legal terms in order to claim a particular identity. Who is Chinese? Who is Hong Kongese? How does one define Chineseness? The films do not provide direct answers to such questions, but offer some hints. In *Comrades,* the cultural and national identity of the protagonists Li Qiao (Maggie Cheung) and Xiaojun (Li Ming) is grounded primarily in their common love of the songs of the Taiwanese singer Deng Lijun (Teresa Teng). Toward the end of the film, the news of the singer's death is broadcast on TV, and the broadcaster says, "Some people say: wherever there are Chinese, Deng Lijun's songs can be heard." In recognizing that cultural productions like songs can become strong markers of identity, the film posits identity formation as a transnational, deterritorialized activity. For the characters in *Comrades,* Chineseness does not necessitate an identification with a specific national and political entity, but can be constructed via something as simple as a song. Perhaps this flexible handling of identity in *Comrades* should not come as a surprise, since, after all, it is only logical that a territory handed off from one nation to another would spawn films in which nationality is borderless, a state of mind rather than of place.

It appears in historical retrospect that Hong Kong films dealing with the theme of identity and diaspora have responded to three key defining moments in the 1980s and 1990s: the Sino–British Joint Declaration in 1984, the Tiananmen incident in 1989, and the handover in 1997. These decisive events have gone a long way toward creating the "collective unconscious" of Hong Kong residents, as it were. The residents of Hong

Kong knew for sure that the island would return to mainland China after Deng Xiaopeng and Margaret Thatcher concluded the Joint Declaration in 1984. Hence, Hong Kong cinema looked to the mainland as the motherland, the root, the home, and the "father" with nostalgia and attachment. In response, Hong Kong film from the period exhibits a "China Syndrome," in which China becomes an object of fascination and longing (a prime example is Yim Ho's *Homecoming* of 1984). A few years later, the Tiananmen incident in spring 1989 shattered the hope of a happy unification with the mainland. As a result, Hong Kong films began to express a feeling of repulsion against the mainland and delineate the experiences of alienation, homelessness, displacement, and exile on the part of Chinese nationals and Hong Kong residents. Examples of this second wave include Ann Hui's *Song of Exile* (1990), Evans Chan's *Wrong Love* (aka *Crossings*, 1992), Clara Law's *Farewell China* (1990), and Stanley Kwan's *Full Moon in New York* (1990). Since 1992, the Chinese government has officially adopted the principle of a "socialist market economy," and Hong Kong has been the biggest foreign investor in mainland China. Its capital propels the economic development of China, which in turn fosters a prosperous Hong Kong economy and makes its business tycoons grow richer and richer day by day. The memory of the Tiananmen incident dims in the minds of business leaders and politicians in Hong Kong, China, and the world. Transnational capitalism is the *modus operandi* of daily business. Thus, a flexible and transnational representation of national and cultural affiliation has arisen in Hong Kong films as well, especially in Wong Kar-wai's films.[22] In these third-wave films, Hong Kongese identity is not necessarily defined in relation to China as a nation-state, but often happens in diasporic, transnational settings. This filmic fluidity of identity is echoed in the real lives of a number of high-profile Hong Kong film talents, who operate truly across the borders of the film and TV industries of both East and West: these talents include John Woo, Jackie Chan, Chow Yun-fat, Michelle Yeoh, Maggie Cheung, Jet Li, Tsui Hark, Ringo Lam, Yuen Woo-ping, Sammo Hung, and others. Their hybrid film productions and performances have become the sources and materials of popular entertainment for audiences in Hong Kong, the West, and all over the world.

The end of colonial rule in Hong Kong and its return to China prompted renewed reflections on the nature of China and the Chinese not just in Hong Kong, but across the globe. As I mentioned earlier, the Hollywood and mainland films I have discussed all purport to tell big stories about China. Their point of departure is to examine China as a geopolitical entity, a nation-state, rather than to look at the question of identity formation as a flexible, fluid process that happens at the level of the individual. Inevitably, Hollywood and mainland Chinese films reveal an embedded ideological and political opposition between the East and the West. The films from both countries tend to fall back on cultural stereotypes and national animosities. In the productions, portrayals of China and Chineseness seem to be vehicles for the East and the West to work out their anxieties over China as a nation-state in particular and to explore more abstract questions about nationalities and borders in general. We should probably not therefore expect to see genuinely "borderless" films from either Hollywood or mainland China anytime soon—at least not before both countries take much greater steps toward a transnational perspective themselves. In contrast, films from Hong Kong display a much more supple touch. Instead of invoking grand historical narratives or relying on rigid political positions, Hong Kong films explore the possibility of a thoroughly transnational ethos.

Notes

1. Earlier versions of the essay were presented at the Asia Society in New York in December 1998; the annual convention of the Association for Asian Studies in March 1999; the Centre of Asian Studies at the University of Hong Kong; the Department of English at the Chinese University of Hong Kong; and the conference "Film, Television, Media, and Popular Culture in Taiwan, Mainland China, and Hong Kong" in Taipei, Taiwan, in summer 1999. I am grateful to Robert Radtke, Elizabeth Sinn, Kwok-kan Tam, and Lee Tain-Dow for inviting me to speak on these occasions. Robert Radtke kindly sent me relevant critical materials about the Hollywood films under discussion.
2. Roland Robertson, *Globalization: Social Theory and Global Culture* (London: Sage Publications, 1992), 9.
3. John Fiske, "Global, National, Local? Some Problems of Culture in a Postmodern World," *The Velvet Light Trap* 40 (1997): 57.
4. Alan Riding, "Why *Titanic* Conquered the World," *New York Times* (April 26, 1998), section 2:28.
5. Mike Featherstone, "Localism, Globalism, and Cultural Identity," in *Global/Local: Cultural Production and the Transnational Imaginary,* ed. Rob Wilson and Wimal Dissanayake (Durham: Duke University Press, 1996), 58. For an outline of several major theories of modernity in the contemporary Chinese context, see my essay "Universality/Difference: The Discourses of Chinese Modernity, Postmodernity, and Postcoloniality," *Journal of Asian Pacific Communication* 9 (1999): 97–111.
6. Roger Ebert, review of *Red Corner, Chicago Sun-Times* (October 1997); website—http://www.suntimes.com/ebert/ebert-reviews/1997/10/103102.html.
7. Kenneth Turan, "*Corner:* A Heavy-Handed Battle with Justice in China," *Los Angeles Times* (October 31, 1997).

8. Janet Maslin, "*Red Corner:* Melodrama-cum-Credibility Snag," *New York Times* (October 31, 1997).

9. Roger Ebert, review of *Kundun, Chicago Sun-Times* (January 1998); website—http://www.suntimes.com/ebert/ebert-reviews/1998/01/011604.html

10. Stephen Holden, "*Kundun:* The Dalai Lama, Toddler to Grown Man, in Exile," *New York Times* (December 24, 1997).

11. Kenneth Turan, "*Kundun* Lacks a Certain Presence," *Los Angeles Times* (December 24, 1997).

12. Roger Ebert, review of *Seven Years in Tibet, Chicago Sun-Times* (October 1997); website—http://www.suntimes.com/ebert/ebert-reviews/1997/10/101003.html

13. Kenneth Turan, "More Pitt than Politics," *Los Angeles Times* (October 8, 1997).

14. Kevin Thomas, "Lovingly Wrapped *Chinese Box* Transcends Melodrama," *Los Angeles Times* (April 17, 1998).

15. *Tomorrow Never Dies,* a 1997 installment of the endless cycle of James Bond films, also takes up the relationship between China and the West but adopts a new strategy in East-West geopolitics. The film tells a story of collaboration and potential romance between a British and a Chinese secret service agent. For a detailed study, see Anne T. Ciecko and Sheldon H. Lu, "The Heroic Trio: Anita Mui, Maggie Cheung, and Michele Yeoh—Self-Reflexivity and the Globalization of the Hong Kong Action Heroine," *Post Script* 19 (1999): 70–86, esp. 78–81.

16. See, for example, the film ad in the *New York Times* (May 8, 1999), AR18.

17. Richard Corliss, "Once the transpacific princess of good films and bad, Joan Chen is now an award-winning auteur," *Time* (April 5, 1999), 61.

18. Orville Schell, "Once a Shangri-La Where China Now Dominates," *New York Times* (May 9, 1999), AR31.

19. I thank Shelly Kraicer for finding the relevant information for me about this American actor.

20. For relevant discussions of the film, see Pu Feng, ed., *1997: Xianggang dianying huigu* (*1997: Retrospect of Hong Kong Cinema*) (Hong Kong: Hong Kong Film Critics Society, 1999), 244–47.

21. For a detailed study of the two films, see Sheldon H. Lu, "Filming Diaspora and Identity: Hong Kong and 1997," in *The Cinema of Hong Kong: History, Form, Genre,* ed. David Desser and Poshek Fu (London and New York: Cambridge University Press, 2000).

22. See my "Filming Diaspora and Identity"; see also Yeh Yueh-yu, "A Life of Its Own: Musical Discourses in Wong Kar-wai's Films," *Post Script* 19 (1999): 120–36.

Chapter 7

The Making of a Revolutionary Stage: Chinese Model Theatre and Its Western Influences

Xiaomei Chen

Until recently, literary and cultural critics of Chinese theatre would never have imagined in their wildest dreams that the study of "modern revolutionary model plays" (*geming yangbanxi*)—the only form of literature and art officially promoted during the ten years of the Cultural Revolution in China—could attain significance comparable to that associated with the study of Shakespearean plays, which obtained prominence even in non-Western countries. As those countries formed their own literary traditions, they looked toward the Occident—and therefore, by definition, toward the canon—for literary and artistic criteria. When China, from 1980 to 1982, hastened to resume the production of Shakespearean plays after the death of Mao, it saw justification in the high art of the Renaissance for denouncing the low art, the non-art, or the pseudo-art of the revolutionary model theatre, which was devalued as political propaganda of the worst kind.[1] China's obsession with Shakespeare at a time of national crisis and of new nation/state building in early post-Mao China presents an interesting reversal of what Stephen Greenblatt terms "marvelous possession," the European stratagem of co-opting non-European peoples by taking possession of their properties, which Greenblatt identifies as a feature of the Age of Discovery.[2] In the Chinese case, Shakespearean plays and the aesthetic values they supposedly embodied, that is, the "wonder of the old world," stimulated the revival of the Chinese people's own culture in the post-Mao era.

Surprisingly, there are a number of similarities between model theatre—which chronicles a grand and glorious transitional period in Chinese history—and Shakespearean history plays, which construct a national past of service to the pressing needs of the Elizabethan era. Although Leonard Tennenhouse has argued that dominant aesthetic discourses have prevented Anglo-American literary critics from recognizing the political features of Renaissance writing,[3] the opposite is the case in China. There, the assumption of Shakespearean plays' artistic superiority has paradoxically justified numerous ways of reading Shakespeare politically and ideologically, ways that are themselves mainstream approaches to literary studies in the People's Republic of China validated by Marxist and Maoist theories of literature and art. In early post-Mao Chinese theatre, for instance, some Chinese audiences could not help but reflect on their own traumatic experience during the Cultural Revolution while watching what were deemed highly aesthetic Renaissance plays. Shakespeare's *Macbeth* was interpreted as reminiscent of the tragic story of Mao Zedong, whose power-driven wife was to blame for having coerced him into intrigues and murder. Similarly, *The Merchant of Venice* was understood politically as dramatizing the role of competition in a market economy; hence, Shakespeare's Shylock was seen as promoting the idea of commercial efficiency in a new era of economic reform in contemporary China.

The cultural and ideological dynamics of Cultural Revolutionary theatre shaped—and was shaped by—the political contingencies of China in the 1960s and 1970s. As a powerful cultural memento, model theatre tells us a great deal about the way a people and a nation envisioned the self, imagined the other, and, as a result of coming to an understanding of the other, mirrored the self. In many different ways, model theatre did not simply reflect the cultural and ideological dynamics of the period that gave rise to it, but instead significantly contributed to the ways in which China imagined itself in the public arena of national and international drama.

The model theatre that was promoted during the Cultural Revolution was a theatrical means of evoking the Maoist memory of a past revolution and, with its re-creation on stage, became a measure for realizing a continued revolution in post-1949 China. Importantly, although model theatre is in many ways entirely indigenous to China—a dramatic form that arose in response to specific historical and social forces during the Cultural Revolution—it is also, although much less obviously, indebted to Western traditions, depending for its success on genres, media, and techniques imported from the West. Surprisingly, this highly political and therefore anti-Western form of drama could not have flourished without its Western features. In fact, Occidental influences played as great a role in spurring the development of model theatre as did Chinese traditions.

During the first three years of the Cultural Revolution, from 1966 to 1969, when schools, libraries, and all other cultural institutions were closed in China, the purveyors of the official Maoist ideology promoted what were then known as the eight revolutionary model plays, which consisted of five Beijing operas, two modern ballets, and one symphonic work, with the last three exploring artistic forms imported from the West.[4] At the height of the Cultural Revolution, almost everyone was compelled to see these plays for the sake, so it was said, of each citizen's political education; sometimes performances even preceded or came at the end of political meetings. One of the most important subjects of this model theatre was the enactment of a past revolution, which, as we have seen, was presumed to spur the Chinese people to a continued revolution, the Great Proletarian Cultural Revolution. Paradoxically, it was the goal of this continued revolution to eliminate the very group of top Communist Party leaders who had made the 1949 revolution possible.[5]

Of the eight revolutionary model works, five were direct representations of the revolutionary war experience. The subject of the Beijing opera *Shajiabang* is an armed struggle during the Anti-Japanese War in which Guo Jianguang, a political instructor of the New Fourth Army, and seventeen wounded soldiers defeat Kuomindang troops who collaborated with Japanese invaders.[6] The Beijing opera *Raid on the White Tiger Regiment (Qixi Baihutuan)* depicts the War to Resist US Aggression and Aid Korea (1950–53), during which Yan Weicai, leader of a scout platoon of the Chinese People's Volunteers, overthrows the invincible White Tiger Regiment of the South Korean army advised by American military personnel.[7] In the Beijing opera *Taking the Bandits' Stronghold (Zhiqu weihushan)*, Yang Zirong, a People's Liberation Army scout, ventures into enemy headquarters disguised as a bandit in an effort to liberate the poor people from the mountain area in the northeast during the War of Liberation.[8] The revolutionary modern ballet *(geming xiandai wuju) The Red Detachment of Women (Hongse niangzijun)* has Wu Qinghua, a peasant girl who fled enslavement by a local tyrant on Hainan Island during the Second Revolutionary Civil War (1927–47), joining a women's detachment that is fighting Kuomindang soldiers.[9] The other model works, although they are not primarily war stories, have narratives that evince some significant effects of the wartime period in which they are set. In the Beijing opera *The Red Lantern (Hongdeng ji)*, for example, which is set in the Anti-Japanese War period, an important facet of the plot concerns Li Yuhe and his family's successful transmission of secret codes to the guerrillas in the northern mountains. Similarly, in the modern revolutionary ballet *White-Haired Girl (Baimao nü)*, which is also set in the Anti-Japanese War period, Xi'er, a much exploited subaltern woman, is rescued by the revolutionary army on its way to the battlefield to repel Japanese invaders.[10]

The Chinese ruling ideology at this time derived many benefits from reenacting a past revolution on stage. The plays and ballets served to divert the attention of the populace from their severe poverty, an intractable problem that defied economic solution and that, hence, the government tried to obscure and dismiss from public scrutiny. Served up as spiritual sustenance, the model theatre charged the masses with the revolutionary energy required to defend the fruits of a hard-won revolution in the past. Thus theatre functioned as a shelter from the chaos and civil war that prevailed at the peak of the Cultural Revolution. However, by assembling on stage oppressed workers and peasants who were depicted as a new class of proletarian men and women, theatre in the long run intensified the tension between the different social groups. For when the proletariat was portrayed as the backbone of revolution—both theatrically on stage and theoretically in the pages of Chairman Mao's red *Quotations* book—the myth of the past revolution having created a world of equality was completely demystified.

To explain how he had achieved the national revolution from which the People's Republic was born in 1949, Mao had once alluded to poverty as the source of revolutionary energy: "The ruthless economic exploitation and political oppression of the peasants by the landlord class forced them into numerous uprisings against its rule. . . . It was the class struggle of the peasants, the peasant uprisings and peasant wars that constituted the real motive force of historical development in Chinese feudal society."[11] This statement of Mao's was frequently quoted in the official reviews of the model theatre, which was said to have dramatized the miserable lives of the poor people before 1949. Yet in post-1949 China, the majority of the people, especially those in the countryside, were still poverty-stricken. They were poor before 1949, after 1949, and perhaps poorer still during the Cultural Revolution, when a civil war disrupted normal agricultural activities. That this mass poverty did not lead to another revolution was a function of the severing of the link between revolution and an appropriate ideology—a *sine qua non* for the successful sequence from poverty to revolution—as so constructed by the Maoist ideology.

This disruption was exemplified in the model theatre. Through the cultural practice of "recalling the bitter past in order to appreciate the happy present (*yiku sitian*)," the roots of poverty were conveniently traced to the old society, which "forced a human being to turn into a ghost."[12] Glorified by contrast, the new society presumably "changed a ghost back into a human being again." In the model ballet *White-Haired Girl,* for example, the poor peasant girl Xi'er, having escaped from Huang Shiren, the vicious landlord who had had her father beaten to death for failure to repay his debts, survives in a mountain cave as a wild person, her hair finally turning white. After being rescued from her ghostly existence by the arrival of the Communist army, she joins the revolution so that she can follow Chairman Mao's example and liberate other poor sisters like herself.[13]

Chinese audiences attending this ballet were supposed to compare Xi'er's ghostly life with contemporary life in socialist China, which was supposedly infinitely better. Official press articles documented such reactions from members of the audience. Li Shanyuan, a "model peasant"[14] from Lugouqiao People's Commune, for instance, was quoted as saying: "When I saw [Yang Bailao] beaten to death and the vicious way [Huang Shiren's] family treated [Xi'er], I thought: That isn't make-believe; this is [exactly] what poor peasants had to put up with in the old society."[15] What was being accomplished here was the replacement of the theatrical convention, suspension of disbelief, with that of suspension of political belief. As Li's additional comments indicate, the testimony from survivors of the old China was meant to valorize both past and present class distinctions: "This ballet teaches us not to forget the crimes of the landlord class, not to forget class hatred; it teaches us that it was not easy for us people to win mastery over our land and we must keep the state power firmly in our own hands."[16] Many statements also were obtained from female audience members who claimed they had been subjected to similar exploitation in the old society, or had known someone who had gone through the same kind of experience. It thus appeared that the female agency onstage was speaking for the subaltern communities offstage, whose stories about their own hardships caused by the staged events then served to reinforce the legitimacy of the official ideology.

Similar statements surfaced from members of the cast and were also printed in the official press. Ling Kue-ming, who, in *White-Haired Girl*, danced the part of Wang Dachun—the Communist soldier who saves Xi'er from a life in the wild mountains—gave his own account of the bitterness of the old society. Born into a poor peasant family, "I could never have attended a primary school," Ling said, "to say nothing of going to a school of dancing."[17] In the old society, in order to escape the oppression and exploitation of the landlord, Ling claimed, his family went to Shanghai from the countryside to escape poverty. Still unable to feed his family, his father had to send Ling's elder brother to an orphanage. The elder brother's tragic departure flashed upon Ling's mind, thus enabling him to express true feelings through his performance.[18]

These comments and their impact on Chinese audiences illustrate one of the chief ironic functions of model theatre: whereas inside the theatre a revolution was being rehearsed night after night, outside the theatre another sorely needed revolution against the ruling class was being fended off in part because of the dramatization of the past revolution inside the theatre. The inside and outside theatres, together with their respective, professed illusory and real worlds, fundamentally contradicted each other, and yet, by doing so, significantly reinforced each other. Thus at this moment in Chinese history, it was theatre, and theatre alone, that intervened in, contributed to, but then severed the potential link between cause and effect in the revolutionary process. Theatre participated in the writing of modern Chinese history, which, as Jiwei Ci points out, "serves as a record of debts, the debts owed by the ruled to the rulers," with the expectation on the part of ruling ideology that each and every reading of past bitterness will "renew and redouble the memory of debts and the readiness to act as debtors."[19]

The history of *White-Haired Girl* presents a number of interesting additional ironies. The ballet version of the work was adapted from an earlier folk opera (*geju*) premiered in Yan'an, the capital of the Communist-liberated area, in April 1945, on the occasion of the Seventh Chinese Communist Party Congress. It was warmly received by Mao Zedong, Zhou Enlai, Zhu De, and other party representatives, and Mao was even quoted as suggesting that Huang Shiren, the vicious landlord, be executed at the conclusion of the opera.[20] According to He Jingzhi, the main scriptwriter, the opera was based on popular folklore about a "white-haired goddess" who was forced by a despotic landlord to repair to the wild mountains. Because she was a "white-haired girl," she was mistakenly worshipped as a goddess and lived on the food offered to her by the local people in a temple.

With its foregrounded theme of the life-and-death struggle between the rich and the poor, the folk opera was enthusiastically received in the liberated areas during the 1940s revolutionary war period. Consequently, it was often performed at mass meetings in order to raise the peasants' class consciousness and turn them against such class enemies as Huang Shiren. The folk opera's success was frequently attributed to the implementation of the principles laid down in Mao Zedong's *Talks at the Yan'an Forum on Literature and Art* (published in 1942), in which Mao demanded that literature and art serve the interest of the people and that it accomplish the central task of the Chinese revolution. Indeed, as the following report indicates, the popularity of the folk opera *White-Haired Girl* seemed unprecedented: Wherever the red flags fluttered, the opera would be successfully performed and loud cries for the revenge of Xi'er would be heard everywhere.

Yet this piece of revolutionary theatre was paradoxically attacked during the Cultural Revolution as a "counterrevolutionary" work that had presumably been influenced by Liu Shaoqi's bourgeois approach to literature and art. The opera was accused, among other things, of portraying "*zhongjian renwu*," or "middle-of-the-road characters," who were neither positive (proletariat heroes) or negative (bourgeois or counterrevolutionary). Yang Bailao, the attackers maintained, took his own life (after having been forced to sell his daughter Xi'er to Huang Shiren) without rebelling or even raising any protest against the injustices he suffered, thus distorting the heroic image of the revolutionary peasant. In contrast, argued leftist radicals, the ballet version of *White-Haired Girl* promoted during the Cultural Revolution more accurately depicted the historical truth; it portrayed Yang as filled with the spirit of revolt and beaten to death by Huang Shiren after a heroic fight. The ballet revision also contributed a heroic and revolutionary Xi'er, unlike the Xi'er in the original folk opera, who had been characterized as so naive that she fantasized about marrying Huang Shiren after having been raped by him and giving birth to his child. This and other plot elements in the opera were criticized as attempts to deliberately obfuscate the class conflicts between the poor peasants and their oppressors.[21] To foreground the correct political stance of the ballet, an epilogue was added in which Xi'er and other villagers join the revolutionary army to demonstrate their infinite gratitude to their party and their love for their great leader, Chairman Mao, a feeling echoed in the theme song at the conclusion of the ballet: "Mao Zedong is the Sun, / The Sun is the Communist Party."[22] One of the main achievements of the ballet, attributed to Comrade Jiang Qing, or Madam Mao, was said to be its emphasis on armed struggle, a key issue that divided Mao's revolutionaries from Liu Shaoqi's counter-revolutionaries in the party's history.

The crowning irony in the history of this work was that in the post-Cultural Revolutionary discourse, when Jiang Qing was being denounced as the archenemy of the people, the past verdict of the folk opera *White-Haired Girl* was reversed. Now its past condemnation was frequently cited as part of the plot by the members of the Gang of Four to subvert Mao's correct approach to literature and art. Jiang Qing's promotion of model theatre—of which the ballet version of *White-Haired-Girl* was a prime example—was further denounced as a classic instance of using theatre for antiparty activities, an accusation previously leveled at the opponents of the Gang of Four. Perhaps no other chapter in history reveals such an intimate and ironic relationship between theatre and politics, between the fate of a revolution and the metamorphosis of that revolution's repercussions in theatrical representations.

The divergent representations of revolution in theatre reconstructed the past history of the party and of modern China during the Cultural Revolution, through revised plays of model theatre. The dual purpose of these different representations was to illustrate the correctness and invincibility of Mao Zedong Thought and to consolidate the power of Mao. Together with the much acclaimed publication of the revised edition of Mao's *Collected Works* and the revised edition of the party's history book—which deleted the undesirable names and events that appeared in earlier editions but were now considered counterrevolutionary—the revised

model theatre became a history maker that heralded a new age by rejecting the old. Of course, *White-Haired Girl* is far from being the only work
whose history is closely associated with the rewriting of China's political
history. The history of almost every other piece of model theatre exhibits
similar revisions and diametrically opposed interpretations as a consequence of changing political conditions and shifting ideologies.

The richness of model theatre as a powerful cultural memento, however, transcends its placement at the center of political discourse. Moving
beyond political analysis, recent scholarship has explored the complexities
and paradoxes of early versions of model theatre to highlight the convergence of diverse discourses already embedded in the texts. In an exemplary
study of this kind, the essay "*White-Haired Girl* and the Historical Complexities of Yan'an Literature" by Yue Meng, the author demonstrates how
the evolution of the different versions of *White-Haired Girl* constitutes a
typical process in which the discourses of *xin wenhua* (new culture), *tongsu
wenhua* (popular and folk cultures), and *xinde zhengzhi quanwei* (newly established political authority) are woven into one text that meshes the traditional cultural values with the modern and contemporary.[23] According
to Meng, the folklore tale of the white-haired goddess was first collected
by men of letters in the Yan'an liberated area as part of their efforts to learn
and adapt material from the regional and popular cultures while creating
their own new, revolutionary culture. At the same time, many of these
writers were heavily influenced by the new cultural movement arising
from the May Fourth period, which, in turn, was receptive to Western influences in the arts, such as that of the European operatic tradition. The
most immediate impact, of course, was that exerted by the political culture
and ideological circumstances of the liberated areas, where land-reform
and thought-reform movements dictated much of the content of literary
production.[24] The result of these concurring influences, the folk opera
White-Haired Girl (1945) is a multilayered text that combines elements of
the old and new cultures, the foreign and indigenous cultures, and the
urban and rural cultures that prevailed until the mid-1940s.[25]

Meng's detailed analysis of the folk opera version demonstrates that although the political concern with the way a new society turns a ghost
back into a human being supposedly touched the hearts of millions of
people in and out of the liberated area, traditional aesthetic and ethical
principles played at least equally important roles. The tragedy surrounding
a father's sudden death on New Year's Eve, the desolate girl seized from her
fiancé just before the wedding, the lovers' hopeless separation followed by
their tearful reunion, and the revenge taken on a vicious man in the end—
all these are familiar narrative structures in Chinese popular culture and
folklore. Long before he had been branded archenemy of the poor people

in a Marxist context, Huang Shiren, the rich landlord, had already been tagged archenemy by the ethics of the popular culture. Similarly, when Dachun, Xi'er's fiancé, returns to his home village, he is recognized first as the local boy whom everyone trusts before his new political status as a Communist soldier is accepted. In both instances, political legitimacy rests on such nonpolitical foundations as ethical principles and community practices.[26]

Meng further argues that the film version of White-Haired Girl (1950) translates the ethically familiar tale of beihuan lihe (the sorrows and joys of partings and reunions) into a popular love story. Although the latter, with the aid of such cinematic techniques as the close-up to accentuate Xi'er and Dachun's passion and the agony of their separation, was designed to entertain urban dwellers, it was still the traditional love story that legitimized the political discourse of class oppression and struggle contained in the film. It is important to note that the film version cut out the scene in which Xi'er fantasizes about marrying Huang Shiren after becoming pregnant with his child. Meng believes the change, intended to emphasize Xi'er's loyalty to Dachun, both strengthens the love story and advances the Communist ideology of class distinction, which prohibits love affairs across class lines. The prominence given to romantic emotions also goes a long way toward transforming the film version into a popular urban text.[27]

The ballet version of White-Haired Girl (1966) was drastically revised in that much of the traditional narrative structure and the popular love story were taken out in order to emphasize the class struggle theme. Both Yang Bailao and Xi'er are portrayed as having a highly developed political consciousness and waging a fearless, heroic struggle on behalf of the exploited class to which they belong. This overt political text, Meng rightly remarks, points up the conflicts and contradictions between the political and apolitical discourses, which in previous versions had been brilliantly blended. Even in this ballet version, however, the tearful reunion of Dachun and Xi'er retains much of the overflowing emotions the lovers exhibited in the earlier versions, thus providing audiences some opportunity to interpret an official text along nonofficial, and perhaps antiofficial, lines.[28]

With its interpretation of model theatre as an embodiment not only of the political context but also of the conflicts and convergence of various cultural, ethical, and literary traditions, Meng's historically grounded study suggests a new way of taking seriously the study of revolutionary literature in contemporary China. In fact, the same trajectory that she pursues in White-Haired Girl can be traced in other works of model theatre. The revolutionary modern ballet The Red Detachment of Women, for instance, was adapted from the film version of the same title, which premiered in 1961.[29] As the winner of four national film awards (for best movie, best director, best actress, and best supporting actor), the film version was extremely popular among Chinese audiences from all walks of life.[30] This was in spite of the fact that as a result of the political atmosphere created by the Great Leap Forward movement, the love story between Hong Changqing, the male party representative of the women's military unit, and Wu Qunhua,[31] the slave girl he rescues from the clutches of the local despot, was reduced to a minimum.

In an essay written in 1962, however, the scriptwriter Liang Xin provides intriguing information about this aspect of the film. It appears that after the film's premiere, he received a letter from an audience member who congratulated him for not having fallen into the "old trap" of telling a conventional love story, which would have overburdened Hong with too many roles, making him at one and the same time a military commander, a mentor, a comrade-in-arms, and a lover.[32] For his part, Liang disagreed and rather regretted that because of the political pressure to express proletariat rather than romantic feelings, he could not fully develop the love story that had been present in the original script version.[33] Yet if one compares the portions of the original script cited in Liang's essay with the finished film version, one can see that even after a few explicit lines were excised, the romantic story was still clear to anyone familiar with the characteristically subtle ways of expressing love in Chinese culture. In my view, the finished film, capitalizing on cinematic techniques that emphasize facial expressions and body language, has the desirable effect of stimulating the audience's imagination and leaving much room for it to fill in the gaps. In fact, Zhu Xijuan, who played the role of Wu Qunhua, admitted that she saw Qunhua (and played her accordingly) as having already fallen in love with Hong Changqing, since, in her view, Qunhua's capacity to love was at least as great as her capacity to hate her class enemies.[34]

There may be some question about the explicitness of the love story between the main characters in the film, but none at all about the love between a pair of less important characters: Honglian, another woman soldier, is happily married to a soldier and gives birth to a baby girl right on the battlefield. As might be expected, both implicit and explicit love stories were left out of the ballet version of *The Red Detachment of Women* that was given much play during the Cultural Revolution. To the Chinese audience familiar with the horizons of literary expectations[35] provided by way of the film version, the potential love story between the male and female protagonists still loomed large. Furthermore, the film and ballet art forms (both of which, not incidentally, had been imported from the Occident) were equally capable of replacing what had been linguistically omitted. Whereas the cinema had the close-up and other techniques to convey emotions, the ballet had the duet, which permitted body contact between a male and a female dancer and made sexual attraction quite palpable.

Recent literary history gives credit to Premier Zhou Enlai as the first state leader to come up with the idea of using the Western form of ballet to create theatrical works to represent the revolutionary experience for the masses. It is said that in 1963, after watching *The Hunchback of Notre Dame,* performed by the troupe of Beijing Ballet School, Zhou congratulated teachers and students for their outstanding achievements in mastering the Western art form. He urged them, however, not just to re-create characters of prince and fairy, but to experiment with representing revolutionary events such as the Paris Commune and the Russian October Revolution. These Western stories might gain an easier success than Chinese ones, since ballet, after all, might prove too alien to depict Chinese lives and struggles. Enlightened, but not limited, by Zhou's directives, students and teachers from the ballet school decided to be so daring in their experiment that they chose *The Red Detachment of Women* as their first try. It proved a resounding success, after incorporating performing conventions from Chinese folk dance, folk music, and Beijing opera. When it premiered in 1964, Zhou Enlai was reported as having been so touched by the ballet that he had tears in his eyes, apologizing to the cast for being too conservative in his earlier suggestion of first experimenting with a foreign story.[36]

A similar story of converting ballet to serve the interests of proletariat art also occurred in the creation of *White-Haired Girl,* whose cast from Shanghai Dance School was inspired by the success of *The Red Detachment of Women.* Eager to produce their own version of revolutionary ballet in the Chinese style, the cast solicited criticism from Shanghai workers, who failed to understand the scene when Yang Bailao tried to commit suicide. The cast members later realized that it was not just the ballet "body language" that one worker failed to understand. He was also puzzled about why Yang Bailao should commit suicide. "When I was insulted and abused by a landlord, I rebelled and killed two of these bad eggs. That was why I escaped from the countryside to Shanghai before liberation. Yang Bailao should fight back. He cannot simply kill himself." The worker's words helped the cast members to realize the importance of the theme of "class struggle," for which *White-Haired Girl* was later celebrated as a model theatrical work.[37] Such a theme also helped audiences such as workers and peasants understand and appreciate Western ballet and finally accept it as an art capable of expressing their own lives.

Ironically, here was a supposedly highly political—and hence essentially anti-Western—version of model theatre that depended heavily for its success on media, genres, and techniques imported from the Occident. To denigrate model theatre from an artistic standpoint, therefore, would have been to denigrate the aesthetic traditions of the Occident, in relation to which model theatre had always seen itself as diametrically opposed, as in effect the other. Yet the impact of the Occident becomes even more apparent in the context of the popularization of model theatrical works during the peak of the Cultural Revolution. Aided by film adaptations of the stage productions that had been running in a limited number of cities, model works were made available to audiences in the vast expanse of the countryside and other remote regions of China. In short, the media of the Occident played as large a role in promoting and developing model theatre as did the incorporated folklore and popular traditions.

Another unique facet of model theatre was its selection of Beijing opera, one of the most perfected of artistic forms, as the genre to be reformed for its use. So exquisite was this art form that foreign countries imported it to revolutionize their own theatres. As Georges Banu points out, European theatre artists such as Stanislavsky, Craig, Brecht, Piscator, Tairov, and Tretiakov, who had been seeking a theatrical model at a time of artistic crisis in their own cultures, viewed Beijing opera—as represented by Mei Lanfang's performance, which they saw in the Soviet Union in 1935—as "the crystallization of their visionary spirit."[38] Meyerhold, for instance, found in Mei Lanfang's Beijing opera a replacement for the Stanislavskian memory of the actor ("I") with a memory of the stage, which focuses attention on tradition and on an artistic past. He wrote: "In the hour of the Russian theatre's 'great turning point,' Mei Lanfang was a beacon; artists began to see that they could continue their resistance through subterfuge."[39] Ironically, this beacon of classical Chinese art was relit during the Cultural Revolution to illuminate the beacon of Mao Zedong Thought, which was not only to direct the Chinese Cultural Revolution, but also, perhaps more significantly, to guide revolutionary movements in the Third World in the global struggle against the imperialistic superpowers.

The memory of the stage, which Banu regarded as presenting Europeans with an ideal model of theatre, was combined, during the 1960s in China, with the memory of a revolutionary past, which constituted many of the central themes of model theatre. It is interesting to note that this reenactment of revolutionary experience was not merely aimed at Cultural Revolutionary audiences. Performers in the model plays were encouraged to make use both of the Stanislavskian memory of the actor (the "I")—a legacy of Soviet dramatic performance that had dominated Chinese stage since modern times—and of a memory of the collective (the "We"). It was believed that by acting out the roles of the revolutionary characters—that is, by creating the revolutionary other while rejecting the nonrevolutionary self—the actors would be reformed (i.e., shed their bourgeois ideology).

In concentrating on and honing every body movement and perfecting every operatic tune, the players of model theatre were rechanneling their energies toward a revolutionary ideal. Through acrobatics, they achieved strict bodily control, usually to depict battle scenes; the larger purpose was to demonstrate the clash of conflicting ideologies and to testify to the "truth" that those better armed with Mao Zedong Thought were the ones who eventually prevailed, both on the stage and in real life. Thus did the revolutionary heroes and heroines depicted by the players seize the theatrical space, which was symbolized by highly theatrical signs that embodied the spirit of the proletariat and its triumph over the bourgeoisie. Red flags and red stars stood for the entirely new empire of Maoist China, whose aim was to universalize the contemporary world in the same way that Confucius had blueprinted his "central kingdom." One of the immediate consequences of the focus on bringing one's body under control was to demonstrate the necessity for discipline, which China desperately needed during this chaotic period in the Cultural Revolution, when the drive to eliminate the four olds (old ideology, old culture, old practice, and old customs) had brought about the collapsing bodies of the city.[40] The emphasis in model plays on having a vast countryside for the Communist fighters to maneuver in while plotting against their enemies in the besieged city with its drowsy bodies was not accidental; it testified to the significance of the social space created by the theatre, which advocated the values of the Cultural Revolution. The highly stylized movements on stage instigated offstage imitation leading to conformity. Knowledge was being absorbed through a unique theatrical experience that not only itself resulted from imitating "real" revolutionary life, but, most important, provided millions of people with a model for more revolutionary behavior in their own "real" lives. Here was a case, then, not uncommon in those times, of life imitating art rather than the other way around.

As the example of model theatre suggests, a remarkable feature of contemporary China's cultural scene is the extent to which theatre is political and politics has assumed an essential aspect of theatricality. One quick way to recognize the latter point is to perceive contemporary China for what it virtually is: a stage on which is enacted an ongoing political drama that has all the actors scrambling to perform the "right" parts to ensure their political survival. One could also see that despite its apparent Chineseness—as seen in the use of traditional operatic form in several model plays—model theatre could never have achieved its effects without an infusion of Occidental influence and traditions. Even during the height of the Cultural Revolution, when the Occident was the apparent enemy in Maoist ideology, the Occident had infiltrated China in forms such as ballet, film, and symphony, and had already been shrewdly co-opted—even when surface appearances hid its presence.

Notes

1. Central Drama College premiered *Macbeth* (*Makebaisi*) in 1980 in Beijing; China Youth Art Theatre premiered *The Merchant of Venice* (*Weinisi shangren*) in 1981 in Beijing; the director training class of Shanghai Drama College premiered *King Lear* (*Li'erwang*) in 1982 in Shanghai. For receptions of these plays, see Xiaomei Chen, *Occidentalism* (New York: Oxford University Press, 1995), 49–58.

2. Stephen Greenblatt, *Marvelous Possessions: The Wonder of the New World* (Chicago: University of Chicago Press, 1991).

3. Leonard Tennenhouse, "Strategies of State and Political Plays: *A Midsummer Night's Dream, Henry IV, Henry V, Henry VIII,*" in *Political Shakespeare: New Essays in Cultural Materialism,* ed. Jonathan Dollimore and Alan Sinfield (Manchester: Manchester University Press, 1985), 109–126, 126.

4. For a typical introduction to these works in English, which was aimed at promoting the Cultural Revolution and its model theatre to a readership outside of China, see "Magnificent Ode to the Worker, Peasant, and Soldier Heroes," *Chinese Literature* 12 (1968): 107–116.

5. For a historical survey of the major events of the Cultural Revolution, see Yan Jiaqi and Gao Gao, *Zhongguo wenge shinian shi* (*Ten-Year History of the Cultural Revolution*) (Beijing: Zhongguo wenti yanjiu chubanshe, 1986), especially 441–48, for a definitive account of the model theatre and its significance in the Cultural Revolution.

6. The revised version of this opera was first published in *Hongqi* (*Red Flag*) 6 (1970): 8–39, the most authoritative party journal. An English translation can be found in *Chinese Literature* 11 (1970): 3–62.

7. The revised version of the opera was first published in *Hongqi* 11 (1972): 26–54. An English translation can be found in *Chinese Literature* 3 (1973): 3–54.

8. An English translation of this opera can be found in *Chinese Literature* 8 (1967): 129–81.

9. The original Chinese script was first published in *Hongqi* 7 (1970): 35–65.

10. The other two model works not discussed in this paragraph are the symphonic work of *Shajiabang*, adapted from the Beijing opera *Shajiabang,* and the model revolutionary Beijing opera *Haigang* (*On the Docks*), which was set in socialist China after 1949.

11. Stuart R. Shram, *The Thoughts of Chairman Mao Tse-Tung* (London: Library 33 Limited, 1967), 13.

12. These are typical phrases used during the Cultural Revolution to illustrate the political theme of *White-Haired Girl.*

13. *White-Haired Girl,* in eight scenes and with a prologue and epilogue, was collectively written and produced by Shanghai Dance School around 1966.

14. "Model peasant" or "model worker" in socialist China is an honorary title usually awarded to an exemplary person, whose extraordinary contribution to his or her community is set as an example for others to imitate.

15. "Comments on the Ballet the *White-Haired Girl*," *Chinese Literature* 8 (1966): 133–40, 136. I have changed the original spelling of the names into the pinyin system to be consistent with the rest of this study.

16. "Comments," 136.

17. Kuei-ming Ling, "Taking up Arms," *Chinese Literature* 7 (1972): 106–8, 106.

18. Ibid., 107.

19. Jiwei Ci, *Dialectic of the Chinese Revolution* (Stanford: Stanford University Press, 1994), 82.

20. The opera was collectively written by Yan'an Lu Xun Literature and Art Academy, with He Jingzhi, Ding Yi, and Wang Bin as scriptwriters and Ma Ke, Zhang Lu, and Huo Wei as composers. The first edition of the opera was published by Yan'an xinhua shudian in June 1946. For an English translation, see *White-Haired Girl*, trans. Gladys Yang and Yang Hsien-yi (Beijing: Beijing waiwen chubanshe, 1954).

21. For sample articles that condemned the opera during the Cultural Revolution, see "An Artistic Pearl Created in the Fierce Struggle Between the Two-Lines," *Guangming ribao* (*Guangming Daily*) May 19, 1967, and "On the Recreation of the Ballet *White-Haired Girl*," *Renmin ribao* (*People's Daily*) July 11, 1967.

22. Lu-yuan Yu, "The Revolutionary Ballet 'The White-Haired Girl,'" *Chinese Literature* 9 (1968): 58–9.

23. Yue Meng, "Bai Maonü yu Yan'an wenxue de lishi fuzai xing" ("*White-Haired Girl* and the Political Complexities of Yan'an Literature"), *Jintian* (*Today*) 1 (1993):171–88, 172. See also Meng, "Female Images and National Myth," in Tani E. Barlow, ed., *Gender Politics in Modern China* (Durham: Duke University Press, 1993), 118–136.

24. Meng, "Bai Maonü," 173–74.

25. Ibid., 175.

26. Ibid., 180.

27. Ibid., 181–84.

28. Ibid., 187.

29. For a factual account of how Beijing Dance School Experimental Ballet (*Beijing Wudao xuexiao shiyan baleiwutuan*) started to adapt *The Red Detachment of Women* from film version to ballet version in 1963, see Dai Jiafang, *Yangbanxi de fengfeng yuyu* (*The Wind and Rain of Revolutionary Model Theatre*) (Beijing: Zhishi chubanshe, 1995), 93–101.

30. According to an article published in *Renmin ribao*, April 28, 1962, the first film award since 1949, known as *baihua jiang* (One-Hundred-Flower Award), was initiated by the film journal *Dazhong dianying* (*Popular Film*), which received 117,939 ballots from the audience members. *The Red Detachment of Women* received four out of the total fourteen

awards for different genres of films. See *"Hongse Niangzijun huo zuijia gushipian jian"* (*"The Red Detachment of Women* Won the Best Movie Award"), in *Hongse Niangzijun: cong junben dao dianying* (*The Red Detachment of Women: From Script to Film*) (Beijing: Zhongguo dianying chubanshe, 1964), 465.

31. "Wu Qunhua"—a traditional female name, with *qun* meaning "beautiful jade" and *hua* meaning "flower"—was the original name in the film version; it was later changed to "Wu Qinghua" in the model ballet version. With *qing* meaning "clean and pure" and *hua* (with a different tone and different character), "China," the revolutionary overtone of the changed name in the model ballet version was clear to most of the Chinese audiences.

32. Liang Xin, "Renwu, qingjie, aiqing ji qita" ("Character, Plot, Love, and Other Issues"), in *Hongse Niangzijun: cong junben dao dianying,* 228–250, 242.

33. Ibid., 248.

34. Zhu Xijuan, "Cong nünu dao zhanshi" ("From Slave Girl to Warrior"), in *Hongse Niangzijun: cong junben dao dianying,* 305–319, 315.

35. For the notion of "horizons of expectations," see Hans Robert Jauss, *Toward an Aesthetic of Reception,* trans. Timothy Bahti (Minneapolis: University of Minnesota Press, 1982).

36. Dai, *Yangbanxi,* 94–101.

37. Dai, *Yangbanxi,* 102–106.

38. Georges Banu, "Mei Lanfang: A Case Against and a Model for the Occidental Stage," trans. Ella L. Wiswell and June V. Gibson, *Asian Theatre Journal* 3 (1986): 153–78, 153–4.

39. Banu, "Mei Lanfang," 158.

40. Yan and Gao, *Zhongguo wenge,* 66–71.

Part III

Crossing Other Borders:
The Politics of Co-optation

Chapter 8

King Kong in Johannesburg:
Popular Theatre and Political Protest
in 1950s South Africa

Cynthia Erb

In August 1999, the South African musical *Kat and the Kings* had its Broadway debut. The fictional story of a 1950s a cappella doo-wop group struggling for fame, *Kat and the Kings* is set against the backdrop of District Six, an urban district of Cape Town that was allocated to people of mixed race (the group designated as "colored" in South Africa). Throughout the 1950s District Six had a reputation for being both dicey slum and thriving cultural center. It was doubtless the latter urban image—as home to a diverse, productive population—that led the apartheid regime to respond in 1966 by declaring District Six an area henceforth reserved for whites, to be cleared of its nonwhite population and bulldozed.[1] Although *Kat and the Kings* had been both a commercial and a critical success in Cape Town and London, it received a mixed review from *New York Times* critic Ben Brantley, who found its musical comedy approach to the events of the apartheid era disconcerting.[2]

Kat and the Kings draws on a historical dramatic tradition prefigured by the stage musical *King Kong*, which opened in Johannesburg in the winter of 1959. The story of a black boxer named Ezekiel Dhlamini, *King Kong* is set in Sophiatown, the legendary suburb of Johannesburg that was one of the last places where black South Africans could own property during the apartheid era. An urban location known to have given birth to both a thriving black popular culture and to various forms of open political dissent, Sophiatown carried such enormous real and symbolic value for black South Africans that when the apartheid regime finished bulldozing it and removing the last of its nonwhite residents to Soweto in the early 1960s, it was renamed Triomf, Afrikaans for "triumph." The current government changed the name back to Sophiatown in 1996—a symbolic act of the post-apartheid era.[3] (The relationship between the stage musical *King Kong* and the RKO film of the same title is complicated, and will be discussed below.)

Dubbed a "jazz opera" by Harry Bloom, author of the musical's book, *King Kong* set a standard in commercial South African theatre: for the first time the format of the American stage musical was combined with African jazz and popular dance traditions common to Sophiatown's nightlife and street culture in an effort to render a significant moment from the apartheid era. *King Kong* was a major commercial success, selling out all performances during a six-month run in several South African cities and eventually traveling to London, where it ran for a year. Its place in South African theatre history derives from its status as a highly visible early work, in which popular theatre was used as a platform for public critique of the apartheid government. In addition, as the *Kat and the Kings* example suggests, *King Kong* has asserted a lasting influence: prior to *King Kong*, musical theatre in South Africa had consisted of variety revues of the sort found in the vaudeville tradition. *King Kong* is credited with introducing the musical genre's combination of narrative and musical numbers in a fashion that would prove influential, not only for commercial productions created by white theatrical artists, but also for a tradition of black popular theatre that gained momentum in the townships in the 1960s and 1970s.

King Kong also retains its lasting historical place by virtue of circumstances that surrounded its opening in February 1959. That year marked the end of what author Lewis Nkosi would later name "the fabulous decade," when apartheid regulations intensified, but when both cultural and political resistance still seemed possible.[4] The timing of *King Kong* was crucial: in February 1959, Sophiatown, re-created on the stage of a downtown Johannesburg theatre, had in reality been almost completely bulldozed, and was thus only recently consigned to memory. Moreover, the infamous Sharpeville massacre, which launched the even more politically repressive 1960s, was only a year away. Writing in the 1960s from a position of exile, Nkosi states that the wild enthusiasm that greeted *King Kong* was probably not for the show itself so much "as it was applause for an Idea which had been achieved by pooling together resources from both black and white artists in the face of impossible odds."[5] For Nkosi, recalling *King Kong* as the bittersweet culmination of a decade consigned to nostalgic memory, the play stands as a symbol of the artists' and audience's tragic naïveté, for believing the show was the beginning of something, rather than the end: "For so long black and white artists had worked in watertight compartments, in complete isolation, with very little contact or cross-fertilization of ideas. Johannesburg seemed at the time to be on the verge of creating a new and exciting Bohemia."[6]

Contemporary critics of *King Kong* acknowledge its historical importance but point to its political limitations as a black-cast musical largely produced by white liberal artists and directed primarily toward a white liberal elite in Johannesburg. Critics such as Robert Kavanagh and Rob Nixon object to *King Kong*'s status as a theatrical "blockbuster," which hybridized Euro-American genre elements and African jazz traditions for commercial gain.[7] Although *King Kong* now appears as a very limited critique of apartheid, I would argue that as a work of popular hybridity, it furnishes an opportunity to examine timely issues in global cultural studies. The producers of *King Kong* effected a translation, from the syncretic musical traditions of Sophiatown (embodied in different forms of African jazz) to a work of commercial pastiche created primarily for Johannesburg audiences. Because *syncretism* and *pastiche* are related terms, I will define them in the context of African studies.

Karin Barber has shown that Africanists divide African culture into three layers: traditional, popular, and elite.[8] For Barber, popular culture is the most vital yet slippery of the three categories, since its definition resides in what it is not—neither traditional nor elite culture. (Barber treats the other two categories, traditional and elite, as far more clearly delimited.) Barber then maintains that syncretic art is essentially urban-popular art defined in relation to traditional African culture. For example, in an influential study of Sophiatown's popular music, David Coplan has shown that many urban South African musical traditions of the 1950s mixed modern, "foreign" forms, notably black American jazz, with aspects of traditional African musical culture.[9] *King Kong,* scored by black South African composer Todd Matshikiza, mixes syncretic African music forms with production numbers of the type found in the American "showtune" tradition.

In a partial revision of Barber's tripartite classification system, I am introducing the word *pastiche* to define the type of commercial popular work oriented in the opposite direction—away from traditional African culture, and toward elite white traditions of European-South African culture. In these terms, *King Kong* appears as a work of commercial pastiche, known for its adoption of the Euro-American musical format, and yet retaining syncretic impulses in the musical score. In my research, I have often found that pastiche works such as *King Kong* can prove to be complex cases: frequently offering rather limited political critiques, they nevertheless tend to circulate widely, leaving behind rich histories of reception. Using discursive materials generated through *King Kong*'s success, I will revisit the "scene" of its production, partly to examine differences between black and white South African approaches to Western mass culture. At the same time, however, I want to consider the "traffic between" syncretic and pastiche productions: critiques of *King Kong* tend to be dismissive of its Westernized format; and yet in this setting there existed such a highly developed tradition within urban black South African culture for creative incorporation of Western "mass" forms that aspects of *King Kong* were ultimately recycled back into some forms of black South African popular theatre—itself one of the most significant syncretic traditions of the 1960s and 1970s.

In addition, I will argue for the significance of questions of performance, which have generally been elided in existing work on *King Kong* but which have a central place in delimiting the borders between local syncretic and commercial pastiche cultures. Although critiques of *King Kong* stress the control asserted by the show's white producers, writers, choreographers, and financial backers, it was the black South African musicians and performers who, charged with instilling the show with the life and zest of Sophiatown in its glory, offered through their performances a mediation between Sophiatown and Johannesburg, black and white, syncretism and pastiche. Critiques of *King Kong* tend to be production accounts saying little about performance, and indeed a paucity of historical evidence means that any attempt to restore conditions of performance must remain partial and provisional. And yet the musical's reflexive form, which summons attention to issues of fame and showmanship, is significant: the show is so determined to "embody" Sophiatown in its protagonist, local athletic star Ezekiel Dhlamini, that by extension it seems that the show was designed similarly to "embody" the vanished suburb in the performances given by Sophiatown's musical stars. Revisiting the context of *King Kong*'s production affords an opportunity to reexamine the shifting borders between syncretism and pastiche, and to demonstrate the extent to which musical performance can become central to the process of marking out these borders.

Historical Background:
Sophiatown in the 1950s

Nkosi's characterization of the 1950s as the "fabulous decade" defines the moment as a rich period in South African history, evoking historical memory and nostalgia. The years delimiting the decade are crucial: in 1948 the National Party won national elections, running on the slogan "apartheid" (literally "separateness"); the year 1960 witnessed the infamous Sharpeville massacre, when police opened fire on a crowd of black protestors who were resisting pass laws. Sixty-seven people were killed, most of them shot in the back. Sharpeville gained a place in history as symbolic inauguration of the more repressive 1960s.

The decade of the 1950s came to be regarded as a moment when South African government policy shifted gradually from segregation to apartheid, with a battery of laws erected to control the movement and residency of black South Africans. The Registration Act of 1950 forced all citizens to be assigned to one of the established ethnic categories of South Africa. These ethnic categories were deemed arbitrary and artificial, in the sense that their purpose was not so much to define social cohesion as it was to help the government place even more controls over residency. Pass laws were enforced as a means of controlling the movement of black males (and in later years, of black women). The Natives Act, which endowed the government with the authority to clear slums, dated back to 1923, but because it took time to build new housing for nonwhites in areas farther out from the city centers, this law was not really enforced until the 1950s, when systematic bulldozing and forced removal projects proliferated. As we have seen, the clearance of Sophiatown was one of these projects, begun in 1956 and completed in 1960.

If the 1950s marked the growing ferocity of apartheid, the late 1940s and 1950s were simultaneously defined by the increasing organization and militancy of black politics. A major miners' strike took place in August 1946 (supported by the Communist Party). In 1949, a bus boycott designed to protest an increase in fares went on for months. In 1952, the African National Congress (ANC) launched its Defiance Campaign, which drew upon the philosophy of Mahatma Gandhi to create a system of strategic acts of nonviolent protest. A white organization known as the South African Communist Party formed in 1953, and increasingly affiliated itself with the ANC. In 1959, the year of *King Kong's* run, the ANC issued its first call for international sanctions against South Africa.

For the history of South African drama, the 1950s marks a trend toward "township plays," the phrase describing a diverse body of works eschewing focus on traditional rural South Africa in favor of the experience of urbanization. This cultural development reflected historical and economic events: by the 1950s over half of South Africa's black population was urbanized. In the case of Johannesburg, black presence was of two types: migrant mine laborers, who resided temporarily in compounds but whose permanent homes were in rural areas; and permanent residents who claimed the city as home. What emerges is a case in which black urban presence was an economic necessity, as the mines required a cheap labor force, and yet a "problem" for a racist government increasingly asserting control, with the result that the place of black South Africans in the cities was extremely precarious. Historian Robert Ross has argued that the entire apartheid system could be viewed as "driven by attempts to control the numbers and behaviour of Africans within South Africa's cities."[10]

Sophiatown in the 1950s has attracted considerable attention from historians, because the suburb became a site for the production of a rich popular culture that seemed to emerge through the contradictory urban experiences of black South Africans in this period.[11] A number of these elements are important for examining *King Kong:* a new black journalism associated with *Drum* magazine; a musical culture rooted in jazz and other popular music; a nightlife organized around shebeens (small clubs where liquor was illegally consumed by black South Africans); and an emerging culture of gangsters known as *tsotsis* (the urban African pronunciation of "zoot suit"). According to Nixon, the *Drum* writers regarded themselves as permanent urban residents and thus did not identify with the migrant black African experience.[12] They consequently scorned South Africa's traditional rural culture, choosing instead to chronicle the modern urban experience of life in Sophiatown.

Journalists at *Drum* wrote passionately about the culture of the shebeens. Under apartheid, blacks were not permitted to purchase liquor, a situation that fostered an underground distribution network tied to shebeens that were largely owned and operated by black women colloquially known as "shebeen queens." The shebeens gave rise to an entertainment tradition that loosely mixed comedy sketches with local jazz performed by groups such as the Jazz Dazzlers and the Manhattan Brothers. Primarily patronized by black South Africans, shebeens also drew some white visitors to the black townships, and as Nkosi puts it, the clubs thus functioned as crucial locations of interaction between the races: "When a new generation of white South Africans grew up which was prepared to rebel against the narrow confines of colour-bar society, the only place, outside the suburbs, where they could meet young Africans was in the 'shebeens.'"[13]

The shebeens also attracted *tsotsis,* gangsters who were themselves enormously influenced by American mass culture. Fanatical moviegoers, *tsotsis* copied their dress from American magazines such as *Esquire,* and their speech and performance styles from American movies—notably Richard Widmark's performance in *Street With No Name.*[14] They also developed a local language called *tsotsitaal,* which fused elements of Afrikaans, African languages, and American slang expressions taken from the movies. Johannesburg *tsotsitaal* was eventually adopted by urban workers, becoming the language of black South African working-class culture.[15]

In his analysis of Sophiatown in the 1950s, Rob Nixon argues that the *Drum* writers admired the *tsotsis* for self-consciously distancing themselves from rural African traditions, and for using American styles, lingo, and mass entertainment forms to create a new African urban identity—achieved precisely at a moment when the status of black residents within South Africa's cities was being severely threatened by the apartheid government. Because apartheid restrictions made it extremely difficult, if not impossible, for black South Africans to gain access to "serious" culture (opera, classical music, and "serious" theatre), there was a widespread tendency to emphasize cultural forms more readily available to the oppressed black majority, especially movies and popular music. *Tsotsis* made a virtue of the easy availability of Americanized mass entertainment, drawing upon it to invent a new urban style that became influential. In this context, it is tempting to suggest that the *tsotsi* habit of creatively mixing elements of mass culture could function as a kind of expressive opposition to the apartheid government's use of African languages and traditional culture as the basis of its relentless efforts to classify all black Africans into "pure" ethnic categories. (The classification system for black South Africans was organized according to a group of African languages.) Indeed, it is important to stress that in the apartheid era, the government tried to use traditional African culture to impose upon black South Africans forms of identity with which they frequently did not identify. Grasping the role of mass entertainment in Sophiatown at this time thus requires an understanding of historical circumstances specific to the apartheid era.

But it would also be wrong to view Americanized mass entertainment exclusively as an arena for the expressive liberation of black South Africans. As apartheid laws proliferated over the course of the 1950s, black entertainers were forced to become increasingly dependent on white sponsorship if they were to work at all. Robert Kavanagh, who has produced the most complete production analysis of *King Kong*, is particularly critical of the organization that staged the musical, the Union for South African Artists, which was operated through white leadership. Originally established in the early 1950s for the purpose of helping black musicians receive copyright protection for their work, the Union became known for sponsoring activities linked to trade unionism, amateur dramatics, and African music.[16] Among its most successful ventures was a series of Township Jazz concerts featuring various local black musical "stars." Since members of the cast and orchestra for *King Kong* were largely drawn from the Township Jazz roster, Bloom's idea for the show can be seen as a shift from the Union's jazz revue format to a full-blown stage musical. Kavanagh argues that in its combination of European elite artistic methods and financing afforded by white industrial capital (the Union had many ties to the Johannesburg mines), the Union ultimately lost track of its original mission of protecting black performers, becoming increasingly commercial, and an exploiter of black talent in its own right. Pursuing a neo-Marxist mode of analysis, Kavanagh concludes that *King Kong*'s efforts to function as a protest work were bound to fail, since the organization backing the musical was so completely determined by economic and social factors characteristic of the apartheid system.

In providing this brief overview of Sophiatown in the 1950s, I have tried to show that within this context Westernized mass culture occupies a complicated place. On the one hand, as Kavanagh suggests in his critique of *King*

Kong's white sponsorship, commercial popular theatre incorporated Western forms that could be construed as expressively aligned with forms of European–South African control operative within the apartheid system. On the other, for black South Africans prohibited from access to "high" European culture and suffering under a government using traditional African languages and culture as the bases of relentless efforts toward ethnic classification, mass culture—and American mass culture in particular (neither European nor "pure" African)—could furnish expressive material for creating new African styles and identities. Given this set of conditions, perhaps it is not surprising that records of black South African response to *King Kong* suggest ambivalence. For example, Bloke Modisane, who closes his memoir *Blame Me on History* with reference to *King Kong,* offers a rather ironic description of the play as "a showpiece of black-white unity of purpose" that was expected to enjoy a "marathon run."[17] His sarcastic description of the play's commercial success invokes the notion of the "blockbuster." And yet Modisane also admits that he ultimately attended performances of *King Kong* a rather astonishing ten times, drawn again and again to the performance of musical star Miriam Makeba in the role of Joyce. Both Kavanagh and Nixon are critical of *King Kong's* status as stage musical, with the implication that Euro-American genre elements dilute the powerful syncretic effects of Sophiatown jazz. In my own analysis, I will not entirely overturn these critiques, but I do wish to give a closer examination to the nature of the show's synthesis of Western and African elements, and, using black journalism and exile literature, I want to consider the variety of black South African responses to *King Kong's* pastiche elements.

Production Analysis: King Kong

King Kong is based on the life of boxer Ezekiel Dhlamini, who went by the stage name "King Kong." (The use of the name "King Kong," both for the boxer and later for the show, is no doubt linked to the extremely successful international rerelease of the film *King Kong* in the 1950s—a point to which I shall return.) Born in Natal, Dhlamini eventually moved to Sophiatown, where he avoided the mining jobs and menial service positions that were imposed on black Africans, choosing instead to support himself as a gambler. Eventually he drifted into boxing, distinguishing himself with an eccentric, unorthodox style, as well as a talent for self-promotion. Famed for his showmanship, Dhlamini was known for standing on street corners and shadowboxing in a fashion designed to attract the attention of the local press and passersby. He also got into fights outside the ring, spending stretches of time in jail.

Dhlamini eventually became black heavyweight champion of South Africa, but apartheid laws prohibited him from officially facing white challengers. (He reportedly fought a white boxer in a secret nighttime match.) After Dhlamini exhausted the boxing opportunities open to black Africans, his manager persuaded him to drop weight and meet the black middleweight champion. To the surprise of all, Dhlamini lost this match and left boxing, eventually becoming a bouncer in a club frequented by *tsotsis*. As his professional life declined, so did his personal life. Suspicious that his girlfriend Maria was cheating on him, Dhlamini stabbed her to death. At his trial, Dhlamini reportedly demanded his own execution, but the judge sentenced him instead to twelve years hard labor. While working at a labor camp located near a dam, Dhlamini walked into the water and drowned himself.

Dhlamini died in 1957, his life quickly becoming a Sophiatown legend, but it is noteworthy that white and black South African versions of his biography tend to differ. For example, in her book on the production of *King Kong*, Mona Glasser casts Dhlamini's life in "tragic" overtones stressing the manifold ways the apartheid system confined the boxer, sending him into a state of depression that resulted in murder and suicide.[18] The mournful tone of Glasser's account contrasts with reporter Nat Nakasa's, which appeared in *Drum* as the play opened in February 1959. Indeed, *Drum*'s juxtaposition of a promotional photofeature on *King Kong* with Nakasa's extremely oppositional story on Dhlamini's life is quite striking. Everything about Nakasa's biography accentuates Dhlamini's physical power and his constant acts of defiance. Nakasa depicts Dhlamini as a "rough," self-taught boxer whose lack of formal training received compensation in physical size and sheer force of determination. Nakasa characterizes Dhlamini's behavior, both in and out of the ring, as eccentric, implying that the boxer persistently dodged the strictures of apartheid. For example, Nakasa describes Dhlamini's famously eccentric fighting style: "King Kong showed plenty of his unorthodoxy that night. Every time he dropped the humiliated Moloi, he would refuse to go to his neutral corner."[19]

Black South African biographies of Dhlamini also construct the significance of his death in a fashion that contrasts with the tone of Glasser's account. Whereas Glasser emphasizes Dhlamini's final stages of despair, Nakasa depicts the boxer as persistently defiant to the end—openly challenging the police during the arrest for Maria's murder, ordering the judge to sentence him to death, killing himself not so much out of despair, but to escape the isolation and tedium of prison life. In her memoir, vocalist Miriam Makeba recalls that many black South Africans disbelieved the story of Dhlamini's suicide altogether: "He goes to jail, where, somehow, he dies by drowning in a pond of water. Everyone wants to know: How did this great strong man, six-foot four, drown in a little pond of water, even in chains? We all suspect some foul play by the authorities."[20]

Nakasa and Makeba provide differing versions of the "facts" surrounding Dhlamini's final days, and yet the two accounts share a tendency to construct the boxer's life as a sort of politically encoded legend, receptive to various truths about life in the apartheid system. Indeed, the possibility that black South Africans were attracted to Dhlamini's life for its potential to highlight recalcitrance within the apartheid system receives reinforcement in Harry Bloom's assertion that he first came up with the idea for a musical about Dhlamini when he heard a street song about the boxer sung in Johannesburg. He mentions that such popular songs, "often treated with sharp satire," were common in Johannesburg, and often developed around topical events, such as a bus boycott, a riot, or a treason trial.[21] Bloom is referring to a tradition of protest songs common in South Africa, but rather than develop this point, he proceeds to describe his ultimate choice of the stage musical format in an expression that will sound all too familiar: "[*King Kong*] would naturally be a musical, not merely because we wanted to give new opportunities to our concert artists, but because township people live in the idiom of musicals. There is nothing artificial about African people breaking into song and dance, or doing so in chorus."[22]

Criticizing Bloom for this type of racist, paternalistic attitude, Kavanagh objects to the author's introduction of the musical format (defined as part of an elite European tradition), while affirming Matshikiza's authentically African musical contributions. My own view of *King Kong* differs in part from Kavanagh's, in the sense that I contend that white and black audience members may have found the musical's synthesis of Western and African elements effective, though for different reasons. Popular genre elements, such as are found in films and musical theatre, tend to be "open" to history, with a great potential for being renovated and recharged according to historical circumstances present within a given context. As I hope to demonstrate, black South African members of the audience may well have construed *King Kong*'s Westernized genre features in light of the recent, timely incidents of Dhlamini's life and death, coupled with the prospect of Sophiatown's imminent "death." Still, Kavanagh persuasively argues for the significance of specific elements in the musical, which were primarily directed toward black members of the audience. These include Matshikiza's score, which mixed showtunes with forms of African music and dance from Sophiatown, including pennywhistle jazz, kwela dance, gumboot dance, *Tshotsholosa* chant, and African choral performance. Moreover, as Kavanagh observes, although *King Kong* was written primarily in English, Matshikiza made sure that portions of the music featured African languages such as Nguni (an urban conglomerate language of Zulu and Xhosa elements), thus ensuring that at least some moments in the show would privilege the linguistic expertise of black South Africans.[23]

King Kong's status as pastiche work derives largely from its incorporation of the form of musical theatre with a stock of popular conventions, many of which derive from cinema—American film genres in particular.

American cinema is known to have had tremendous impact on Sophia-town's local syncretic culture, so it is surprising that cinematic impulses in the musical have attracted relatively little attention. Because many black South Africans at this time were still relatively unfamiliar with Western theatre (Makeba mentions that *King Kong* was her mother's first play), cinema may have furnished a kind of dramatic repertoire shared by white and black South African members of the audience.[24] In addition, the theatre of South Africa at this time was highly regulated and subject to censorship, so that genre codes common in cinema may have provided a sort of dramatic shorthand, capable of suggesting issues of political consequence, but in a discreet, indirect fashion. Focusing on three of *King Kong*'s representational levels—urban life, crime and violence, and the culture of performance it-self—I will suggest that the cinematic impulses present within the musical, which might otherwise be dismissed as entirely formulaic (and thus "pas-tiche"), potentially intersected with emerging discourses about Sophia-town and the apartheid system itself in a fashion that black South African audience members might have found meaningful.

King Kong's two-act structure is fairly simple. The show features a flashback format criticized as facile at the time, but which appears as one of several cinematic effects marking the production as a whole. The play opens in a township in the present (1959), with a chorus of ordi-nary residents lamenting the hardships suffered at the hands of their em-ployers. The chorus members then recall the life of Dhlamini, and there is a flashback to the boxer himself, exercising out-of-doors, surrounded by trainers and reporters (the latter a reference to the *Drum* writers). In prime form, King drops hints to the press that he has prospects for a match in England. Act 1 also establishes a romantic triangle between King, the shebeen queen Joyce, and the *tsotsi* Lucky. Act 1 is studded with production numbers, such as "Back of the Moon," sung by Makeba (as Joyce) and referring to the name of an actual Sophiatown shebeen. Toward the end of act 1, King gets into a fight with a member of Lucky's gang (based on an actual Sophiatown gang called the Spoilers) and accidentally kills him, leading to a ten-month stint in jail. When he returns from jail, the momentum of his career is broken. Act 2 contin-ues to trace the downward spiral of King's fortunes. Joyce leaves King for Lucky. King loses the middleweight match and is publicly ridiculed. Final moments repeat the last stages of Dhlamini's life—murder of Joyce, imprisonment, and suicide.

The play closes by returning to the chorus, whose members remark upon the township's imminent demolition and the relocation of residents. King himself is then revived for the final curtain: "King comes on stage alone, towering, magnificent in his white track suit. He stands in the cen-ter of the stage, glaring at the audience. Then come the red-suited boxers and the whole crowd of fans and admirers, screaming and cheering, and surround King. Flashlight bulbs burst. . . . All the principals come in one by one, and finally King fetches Joyce."[25]

I have quoted the musical's closing lines, because they condense several elements—a powerful, exoticized black male figure, references to "show-making" and theatricality, and the production of fame by the press (flash-bulbs bursting)—that operate in a reflexive dynamic that characterizes the musical as a whole and that appear to have been at least partially inspired by the film version of *King Kong* (as well as some of the other 1930s Hollywood genres—musicals and newspaper films—that are related to RKO's *King Kong*). In his foreword to the libretto for *King Kong*, Bloom states that his show has nothing to do with the film of the same name—possibly to dodge the charge of racism (for comparing a black boxer to an ape character) or to avoid copyright problems. (RKO's ownership of the name "King Kong" may have been the reason Bloom's show never traveled to the United States.) In fact the musical makes only one brief direct reference to the film character, but it features several visual "quotations"—stage tableaux apparently inspired by scenes in the RKO film. One of these tableaux occurs in a production number called "The Death Song," in which Dhlamini sings of the prospect of death while standing behind a gauze curtain, his wrists bound in manacles ("I have always known / You cannot change your road / Nor load / With the name / King Kong" [95]). The show's last production number, "The Death Song," is embedded in a discussion among members of the township chorus of Dhlamini's trial verdict, imprisonment, and suicide. The staging of this number appears to have been directly inspired by a famous scene in the film *King Kong*, in which King Kong is put in chains and exhibited in a show on a Manhattan stage. Though first released in 1933, the RKO film *King Kong* enjoyed an immensely successful international rerelease in the 1950s, which is known to have made significant cultural impact outside North America (famously inspiring the Japanese film *Godzilla*).[26] Despite his disclaimers, Bloom's choice of the musical's title was almost certainly designed to capitalize on the film's renewed visibility for purposes of commercial gain.

The fact that the musical *King Kong* drew from the film *King Kong* forces the question of why black South Africans would have responded positively to a stage production partially based on dramatic comparisons between a black boxer and a fictional ape character—comparisons that are ultimately racist. There are no easy answers to this question. In her memoir, Makeba, commenting on the origins of Dhlamini's nickname, simply states that "the people" had given the boxer the same name as the "mighty creature" from the film.[27] In a certain sense, it seems that black South Africans at this time were inclined to use the name "King Kong" in a way that canceled the bestial image and instead stressed connotations of sheer force and power in an urban setting (the latter quite apt for Dhlamini's public persona). Nakasa's *Drum* biography had used the name "King Kong" this way, as did Matshikiza's account of the Dhlamini trial: "Then King Kong came out and stood feet astride, hips firm, Gulliver style at the top of the stairs. . . . Is he not majestic, big, strong, tough, undaunted, standing up there making the Spoilers and their matches look like Lilliputians?"[28] The famous film scene of King Kong standing on a New York stage, bound in manacles, is often interpreted as a sort of American slave allegory (i.e., forced transport of an exoticized black character to the United States, to be exhibited in chains)—and this despite the fact that many critics would describe the film as a whole as a racist text. In the South African setting, the "King Kong" name, already connoting the suffering of a black exotic, appears to have been reworked with a difference—accenting the name's connotation of power and rage within the modern urban landscape.

Leaving the question of character comparison aside, I would argue that the South African musical is more effective in drawing inspiration from the RKO film's central structuring device, which juxtaposes King Kong's suffering and romantic tragedy with scenes of the destruction of the modern city. Although the theme of the rural innocent in the corrupt modern city was already so common in South Africa's popular culture that it was eventually given a name (the "Jim-Goes-to-Jo'burg" theme, after the title of a 1950s South African film), this aspect of the musical appears to have been directly drawn from the film version of *King Kong*.[29] Translating the RKO film's central concept, the musical offers a structural "rise and fall" of Dhlamini in juxtaposition with the gradual "death" of the township, which is never named but which everyone in the audience would have recognized as Sophiatown. Moreover, I would contend that black South Africans, possessing local forms of knowledge produced by the direct experience of apartheid, were likely to have been in a better position to connect this aspect of the musical's dramatic structure with an emerging set of elegiac discourses that insistently linked Sophiatown's "death" with personal loss. For example, in November 1959, *Drum* ran a photofeature entitled "Last Days of Sophiatown." The story included urban images such as a full-page photograph of a tiny girl crying, a mass of architectural rubble heaped behind her. The news feature's text rhetorically contrasted the "birth" of Sophiatown fame in African jazz, gangsters, and shebeens, with images of urban death. "Sophiatown is now breathing for the last time. . . . Her people do not like the fact that she is being murdered and I sympathize with them because she was a free city."[30] This theme is even more powerfully conveyed in Modisane's celebrated memoir *Blame Me on History*, which makes a motif of connecting the author's inner despair with the destruction of Sophiatown. The book's dedication reads, "To the memory of my father Joseph who was killed by the Sophiatown which they bulldozed into the dust."[31] The opening line reads, "Something in me died, a piece of me died, with the dying of Sophiatown."[32] Another example involves Modisane's walk to visit his birthplace: "I stood over the ruins of the house where I was born in Bertha Street, and knew that I would never say to my children: this is the house where I was born. . . . all that I can bequeath to them is the debris and the humiliation of defeat, the pain of watching Sophiatown dying all around me, dying by the hand of man."[33]

Although *King Kong* has been criticized as insufficiently political, both during its run and in later years, it is important to acknowledge that in a setting in which theatre practices were subject to extreme regulation and censorship, a sustained, direct assault on apartheid would have led to the immediate banning of the show. In contrast to critics dismissive of *King Kong's* pastiche elements, I would argue that the show's extensive "borrowings" of American popular film conventions may have enabled the writers to get around censorship restrictions while at the same time supporting the emotive amplification of the play's limited number of direct topical references. To return to the case of "The Death Song," for example: accounts from the period indicate that audiences found Nathan Mdledle's performance of Dhlamini especially powerful. (Mdledle, who had been in the group the Manhattan Brothers with Makeba, had reportedly known the boxer personally, and based his performance on a careful impersonation of him.) Significantly, this final number is strategically juxtaposed with an extended conversation among members of the township chorus, who refer to the township's imminent demolition, as well as other topical events. One of the characters receives a letter: "On day aforesaid a truck will be put at your disposal to transport your possessions and family to alternative accommodation in Sunshine Gardens" (92). The line refers to the bizarre euphemistic names given to the areas: Sophiatown residents were forcefully removed to Meadowlands Township in Soweto. Another character replies: "This town is coming down fast now" (92). In this same discussion, another member of the township chorus mentions having to choose between the cost of bus fare and purchase of a Coke to quench thirst. For members of the black audience in particular, reference to the cost of bus fare would have immediately signified the whole issue of bus transport of labor from the black townships into the white city of Johannesburg—an issue that inspired intense political protests throughout the apartheid era. What I am trying to suggest, then, is that the cinematic juxtaposition of black protagonist with modern cityscape, for which the RKO film *King Kong* became famous, became in the South African musical *King Kong* the means toward creating a certain evocative spectacle that bypassed censorship restrictions and yet was potentially capable of boosting the power of the show's limited set of topical, political references.

Kavanagh actually acknowledges the importance of *King Kong*'s political references, but he strongly criticizes Bloom's decision to focus almost exclusively on sensational images of Sophiatown life, with a particular emphasis on gangsterism and shebeen culture. Gangsters were in fact central to Dhlamini's life. Matshikiza relates that the Spoilers gang had been present when Dhlamini murdered Maria, and that they served as witnesses for the Crown during the boxer's trial. Still, the stage production tends to simplify issues of crime and violence common in the life of the township. For example, the character of Joyce seems more to resemble a conventional "gangster's moll" than real shebeen queens of the time, who tended to be middle-aged women who found in the illegal trafficking of alcohol a means toward economic independence and the support of their families. (As a shebeen queen told a *Drum* reporter, "I took it on partly because I was brought up to it, partly to educate my children."[34])

The musical *King Kong* takes up actual figures from the Sophiatown scene—Dhlamini, *tsotsis*, shebeen queens—but then renders these in highly formulaic, "cinematic" fashion, essentially creating a romantic triangle in which the *tsotsi* Lucky and boxer Dhlamini do battle over Joyce, who presides over big production numbers staged in the Back of the Moon. Within this triangulated configuration, however, there is a tendency to parallel the protagonist Dhlamini and the gangster Lucky (both played by former members of the Manhattan Brothers), suggesting a situation in which they compete over Joyce but are ultimately similar in their outlaw status. This is reinforced by a developmental pattern in which, despite his dreams of becoming financially independent and gaining respectability (Dhlamini wants to use his boxing money to buy a truck and start a transport business), the boxer finds it increasingly difficult to avoid entanglement with gangster life, and thus to stay out of jail. He loses this battle and becomes a small-time gangster himself.

Admittedly, a number of possible genre sources could have provided the basis for this parallel structure, but it seems to me that the 1954 film *On the Waterfront* (directed by Elia Kazan) may have been especially influential for this aspect of *King Kong*. Kazan's film was a well-known liberal work of the 1950s, and its depiction of the Marlon Brando character as a boxer/martyr figure dragged down by his gangster brother seems to prefigure a similar dynamic in *King Kong*.[35] But I am less interested in proving the direct influence of Kazan's film on *King Kong* than in maintaining that the Hollywood tradition of gangster and crime films (of which *On the Waterfront* is but one example) supported a sort of merger or overlay of "real," historical *tsotsis* with fictional, dramatic versions of the gangster, in a fashion once again amenable to a context under censorship and yet open to topical readings from black South Africans possessing specific forms of local knowledge. The parallel structure I am describing here, in which an innocent protagonist is insistently compared with a gangster or criminal, was extremely apt for rendering one of the most basic aspects of the apartheid experience.

I made the point earlier that *tsotsis* were admired for their creative urban styles, but it is also important to acknowledge that the tendency among some black South Africans (notably the *Drum* writers) to romanticize *tsotsis* as outlaw figures stemmed from a more essential truth: in apartheid, black South Africans were rendered criminals from the start, and the ever-expanding network of regulations that grew up in the 1950s made it virtually impossible for a member of the oppressed black majority to live a life without venturing outside the law. Although this would become one of the most obvious effects of apartheid, discourses for articulating this perspective were arguably still in the process of emerging as late as the year of *King Kong*'s staging—hence, perhaps, the form one finds in a Nakasa *Drum* feature entitled "Criminals Without Crime!," which is illustrated with photographs of ordinary township residents arrested for violating pass laws, liquor offenses, and the like. Statements in the feature suggest that although Africans had been suffering such arrests for years, recognition by the press and the general public was still in the process of emerging: "The police continued their normal routine, arresting thousands of people for petty offences regardless of pleas. . . . It's happening every day in the Union's bigger centers. And it's been that way for years. But lately the newspapers and leading citizens have started saying things about it."[36]

In a context marked by censorship, the dramatic device in which a black protagonist struggles unsuccessfully to disentangle himself from criminal activity arguably played upon a preexistent knowledge among black members of the audience—that victims of apartheid could not move without breaking the law. In a sense then, although production numbers in nightclubs, knife fights among gangsters, and a suffering protagonist following a predetermined path toward imprisonment represent a formulaic repertoire common to the cinematic gangster tradition, these formal codes could well have mixed with emerging topical discourses, with the effect of expressing another dimension of black South African life in apartheid.

I have been concentrating on *King Kong*'s "content," but at this point I want to shift to its staging and performance, for this "metadiegetic" dimension became essential to the show's lasting fame. In a general sense, entertainment venues functioned as significant sites for protest during the apartheid era, because government regulations imposed segregation on both audiences and performers. (During this time, white entertainers could not legally appear onstage with black performers.) A measure of *King Kong*'s lasting fame derived from the white producers' efforts to use the space of the theatre to resist apartheid regulations. Makeba recalls the musical as genuinely transgressive, as the producers insisted on staging the show at a theatre that would admit black South Africans:

> The audiences who come to see *King Kong* are integrated. Mr. Gluckman [the producer] has found a clever way to get around the apartheid laws. The performances are held at the auditorium of the University of Witwatersrand. Since both black and white students attend this university, the authorities cannot tell who are the students and faculty in our audiences and who are people from outside. So people like my mother are seated next to white people who might have been her employers.[37]

In contrast to Makeba's account, which describes audiences as fully integrated, Modisane offers a differing version of seating arrangements for the show. Describing an evening when he attended the show with a white female friend, he notes that she sat in the last row of the "white section" and he sat behind her, in the first row of the "black section"—an arrangement based on segregated seating. It is true that at this time so many major stage productions, including famous works of social commentary such as the imported production of *Look Back in Anger*, were staged at Johannesburg theatres that barred black South Africans from entrance that any effort to stage productions for black theatregoers was deemed progressive; but clearly, Modisane's account indicates that efforts to integrate audiences for *King Kong* were marked by compromise.[38]

In addition to the recognition accorded to the producers' efforts to integrate *King Kong*'s audiences, the musical entertainers' performances (both cast and orchestra members) became important focal points, in part because the musical offered a full roster of well-known Sophiatown stars, but also because it helped launch several of the performers into international fame. The concept of the star admittedly derives from many sources, not only cinema (the area in which I have been concentrating), but also theatre, sport, and musical entertainment. Still, part of the translation from the syncretic to the pastiche that marked the production as a whole can be viewed at a "micro" level through some of the stars—Makeba in particular, who progressed from the Sophiatown jazz scene to a brief appearance in a film documentary about Sophiatown entitled *Come Back, Africa* (directed by Lionel Rogosin) to performances in *King Kong* and eventually to media appearances in the United States, such as a television appearance on *The Steve Allen Show*.[39] Because the presence of the mass media (e.g., newspaper publicity, film and television appearances) was crucial to this articulation of the star, it is arguably the case that discourses of stardom were part of the Western mass-cultural dimension marking *King Kong* as a whole.

Commenting on Sophiatown stars like Makeba, Coplan argues that knowledge surrounding entertainers furnished a kind of public arena for formulating attitudes within apartheid, but he adds that black celebrities often attracted ambivalent responses from black South Africans. On the one hand, it was widely recognized that musical entertainers were constantly forced to submit to white supervision if they wanted to work at all. Coplan notes that "professional performing without white supervision was not considered 'gainful employment,'"—a situation in which musical performers were treated precisely like the unemployed, and were thus subject to especially rigid application of pass laws.[40] On the other hand, performers attracted resentment from black South Africans: "People being strangled by the wider system criticized musicians who played in white nightclubs and City Hall concerts as self-important and sometimes violently attacked them."[41]

In a general sense, star images tend to be composed of quite various, often competing discourses (the better to please a mass audience), and, set against the backdrop of apartheid culture, the contradictory effects of star discourse become especially salient. Part of King Kong's public status as a liberal work stemmed from promotional material that amplified the adversities experienced by its performers, due to apartheid restrictions. Glasser furnishes an anecdote that is representative of the general tenor of publicity for King Kong: she relates that one week into the show's run, one of the leading actors (name not revealed) was arrested by the police for walking out at night, caught without his special King Kong pass. He and a friend were arrested and forced to work in the garden of a policeman's home the next day. The two feared they would be detained so long as to miss that evening's performance, but they were released just in time.[42] Black spectators in particular would have been in a position to bring to the performance a certain preexistent knowledge of the fact that fame and celebrity tended to bring black entertainers "into the light," making them especially vulnerable to some of the apartheid laws (pass laws in particular, because musical performance required frequent traveling, often late at night). Modisane describes the extent to which, prior to her appearance in King Kong, Makeba had experienced considerable harassment from Sophiatown tsotsis, who controlled musical entertainment venues. Black members of the King Kong audience were thus in a position to blend "official" promotional stories about the stars with local forms of knowledge, with the doubled effect of recognizing King Kong as a story about apartheid, performed by persons who had themselves publicly suffered from the effects of apartheid.

This does not change the fact that a star like Makeba ultimately became part of the larger processes of mass commodification: during the months of King Kong's run, Drum featured star-type images of Makeba promoting Coca-Cola (in a full-page ad reading, "Do as the stars do—enjoy delicious, ice cold Coca-Cola") and making an appearance at the Venice Film Festival for the premiere of *Come Back, Africa*.[43] Still, I would argue for the centrality of star discourses for the overall reflexive effects of *King Kong*: the show is about a sports star whose celebrity—which attracts various forms of aggression and negative attention—far from fulfilling his dreams, leads ultimately to his destruction. It seems to me that the use of the star concept as a public image for configuring the way the apartheid system could "derail" one's life and career may have been poignant and resonant for some black South Africans, and this dynamic could well have received reinforcement in the performances of famed local entertainers, known for publicly suffering the effects of apartheid themselves.

As a last example, I want to consider briefly Matshikiza's description of his experiences working on the *King Kong* production, for it resonates with the "double" or reflexive dynamic I have been describing (i.e., that a show about apartheid is put on by persons who are themselves victims of apartheid, and whose very performances bring them more directly into jeopardy from the apartheid system). Matshikiza's account is quite dark in tone, emphasizing the incessant paternalistic behavior of Bloom and the other white creative artists linked with the show as well as a fundamental inability on their part to grasp that creating and producing the show would constantly pose all sorts of risks for the black South African artists. (For example, when Matshikiza insisted that he be invited to creative meetings held at a Johannesburg hotel, he was forced to "pass" as a hotel employee just to enter the premises.) Matshikiza notes that the producers constantly pumped him for information about Dhlamini's life, about Sophiatown, and even about urban African dialect. (This is Matshikiza's sarcastic impression of Bloom's address to him: "I will put some of the language down as spoken in the township, can you give me a few phrases, for instance what do you say when a policeman approaches, what is the lingo?"[44]) Matshikiza also states that Bloom left town during early stages of the writing of the musical, leaving the composer himself and Pat Williams to complete the work: "Black man and white woman caught up in the intrigue of a theatrical project."[45] Although Bloom later took full credit for authorship, Matshikiza's story suggests that in addition to composing the score, parts of the musical's book as well as the "real" knowledge of black life in Sophiatown came from him.

Both Modisane and Matshikiza place discussion of *King Kong* toward the end of their respective memoirs about Sophiatown, thus juxtaposing descriptions of the musical with accounts of their own increasing recognition of the need to flee South Africa for a life in exile. Matshikiza's account begins with a description of his naïve eagerness to cover the story of Dhlamini's trial during his tenure at *Drum*, for he is certain this will be a big break for him, bringing him money and fame. In this account, the courtroom environment ironically becomes the last stage for Dhlamini's legendary showmanship, "the most sensational performance in all of King Kong's ostentatious theater in and out of the boxing ring."[46] As mentioned, Matshikiza tends to stress Dhlamini's "brute force," as if to distance himself, a mission-educated black South African, from the migrant labor class Dhlamini represented. But this becomes part of a rhetorical strategy on the author's part to ironize himself: working in circular fashion, he proceeds from the Dhlamini trial that he believes will bring him fame, through a series of anecdotes about his troubled relationship with the white creative artists associated with the *King Kong* musical, and concludes with an anecdote that would have been extremely familiar to black South Africans at this time. As *King Kong* goes into rehearsal, black cast and crew members must endure the risk of traveling home late at night after rehearsals—a dangerous activity, given curfews and pass laws. One night around midnight, waiting outside the rehearsal hall for the bus chartered to take him home, Matshikiza is approached by Johannesburg police. This causes the composer to panic, for although he carries the special *King Kong* pass, his real papers are not in order. Enduring the mocking and harassment of the police, he decides that very night to go into exile. In this bitter account, Matshikiza quotes the white Johannesburg police who mock him with names like "monkey" and "baboon." Drawing the circle to a close, Matshikiza seems to imply that although he had seen himself as "above" Dhlamini, and had been ready to exploit the boxer's tragedy for his own gain (first as a *Drum* reporter covering the boxer's story, and later as a participant in the creation of the musical based on Dhlamini's life), in the end both boxer and composer endure the same bitter effects of apartheid. Ironically, these effects seem to be made even more acute on account of the celebrity each man had sought.

Conclusion

In this examination of the traffic between local syncretic culture and "mass" pastiche works, my primary purpose has been to complicate the second term. Reading the works of critics such as Kavanagh and Nixon, I have been surprised at clear demarcations drawn, as syncretic culture is affirmed while pastiche is largely dismissed, evidently for its incorporation of Western genre elements. The type of cultural criticism that praises syncretism and criticizes pastiche tends to depend heavily on what I would call "authorship" methods, or methods that stress creative agency. Within these terms, syncretic culture is valuable because it is produced by members of the oppressed group in question (black South Africans in this case). Commercial pastiche culture is dismissed because its creative agents are, for the most part, members of the elite or dominant class. One problem with this approach is that in the case of collaborative productions like *King Kong*, creative agency is more difficult to pin down. The fact that Matshikiza's memoir furiously criticizes *King Kong*'s white creative artists might be deemed sufficient to condemn the whole enterprise, were it not for the fact that the composer leaks information suggesting that his creative contributions far exceeded the musical score, to the extent that he should perhaps be accorded authorial credit along with Bloom and Williams.

In this essay, I have stressed reception methods over the "authorship" approach, and I find that these strongly problematize attempts to separate syncretic from pastiche culture, as well as efforts to praise the former and condemn the latter. When I discovered how frequently discussion of *King Kong* appears in black South African exile literature, I became even more surprised by a widespread critical tendency to dismiss it, partly because I believe commercial works like this, which generate such a range of records of black response, can become "strange" instruments for shifting the production of history itself, as historical accounts of black perspectives are brought to light. But I have also mentioned that black culture in Sophiatown was already heavily based on creative appropriation of Western mass culture, so that it is not terribly surprising that when the black dramatist Gibson Kente saw the show, he essentially reworked its musical format to create a series of black-cast musical productions that became key works within South Africa's black popular theatre of the 1960s and 1970s. In fact, I am dependent on Kavanagh's research for this point, but whereas he tends to stress fundamental differences between *King Kong* and Kente's later musicals, I am more struck by elements they have in common. *King Kong* appears to have provided Kente with a lot of dramatic material: for example, *Too Late* (1975) features a climactic sequence set at a bus rank (seemingly inspired by the use of a bus queue for a major scene in *King Kong*), in which a character dies, and as she falls to the ground, instrumental and vocal music is synchronized with the actions of her fall (a technique similar to movie scoring).[47] In contrast to Kavanagh's emphasis on Kente's dramatic inventions, I would stress his use of conventional dramatic material drawn from *King Kong* and movie traditions. Still, in this essay I have been greatly aided by Kavanagh's research: far from overturning his arguments, I am seeking a shift of emphasis—one that would make it easier to describe an "actual" state of affairs, as pastiche becomes itself the material for the production of syncretic culture.

Before closing, I would add that portions of this essay have actually involved my own interpretive readings of the musical *King Kong*, sometimes running the risk of substituting these for "historic" black responses to the show. Ultimately, however, I am less interested in defending the particulars of these readings than in using them as abstract cases designed to establish a larger point: black South Africans who saw *King Kong* brought to it specific experiences of apartheid and local forms of knowledge (such as *Drum's* journalistic writing) that inevitably inflected their sense of the show's major features. Ignoring this issue, as much of the existing criticism on *King Kong* has done, risks leaving important records of black response to 1950s South African popular culture buried.

Notes

1. Donald G. McNeil Jr., "A Song of South Africa, to a Doo-Wop Beat," *New York Times* (August 8, 1999), sec. 2: 5, 25.
2. Ben Brantley, "Doo-Wopping in Cape Town," *New York Times* (August 20, 1999), sec. B: 1, 5.
3. Hugh Dellios, "Suburb's Name Reminds South Africa of Its Racial Sins," *Chicago Tribune* (December 30, 1996), sec. 1: 1, 12.
4. Lewis Nkosi, "The Fabulous Decade: The Fifties," in *Home and Exile, and Other Selections* (1965; reprint, London: Longman, 1983), 3–29.
5. Ibid., 17.
6. Ibid.
7. Robert Mshengu Kavanagh, *Theatre and Cultural Struggle in South Africa* (London: Zed, 1985), 84–112; Rob Nixon, *Homelands, Harlem, and Hollywood: South African Culture and the World Beyond* (New York: Routledge, 1994), 22.
8. Karin Barber, "Popular Arts in Africa," *African Studies Review* 30, 3 (1987): 9.
9. David Coplan, *In Township Tonight!: South Africa's Black City Music and Theatre* (London: Longman, 1985).
10. Robert Ross, *A Concise History of South Africa* (Cambridge: Cambridge University Press, 1999), 116.
11. For excellent cultural histories of Sophiatown, see Coplan and Nixon.
12. Nixon, *Homelands,* 15–17.
13. Nkosi, "Fabulous Decade," 10–11.
14. Coplan, *In Township Tonight!,* 162.
15. Ibid.
16. Kavanagh, *Theatre and Cultural Struggle,* 84.
17. Bloke Modisane, *Blame Me on History* (1963; reprint, London: Penguin, 1990), 291.
18. Mona Glasser, *King Kong: A Venture in the Theatre* (Cape Town: Norman Howell, 1960), 4.
19. Nathaniel Nakasa, "The Life and Death of King Kong," *Drum,* February 1959, 28. Reprinted in Nathaniel Nakasa, *The World of Nat Nakasa,* ed. Essop Patel (Johannesburg: Ravan, 1985), 121.
20. Miriam Makeba, with James Hall, *Makeba: My Story* (New York: New American, 1987), 68.
21. Harry Bloom, foreword to *King Kong: An African Jazz Opera,* by Harry Bloom and Pat Williams (London: Collins, 1961), 8.
22. Ibid., 11–12.
23. Kavanagh, *Theatre and Cultural Struggle,* 110–12.

24. This is a complicated point: as a mission-educated writer, Modisane had a keen desire for "serious" drama but complains bitterly about apartheid restrictions that largely prevented him from attending white European stage productions. Black South Africans who had not been formally educated at mission schools, such as Makeba's mother, were much less likely to be familiar with Western drama.

25. Bloom and Williams, *King Kong: An African Jazz Opera*, 96. Page references for subsequent quotations from *King Kong* will be provided in the text.

26. For a discussion of the 1950s reissue of *King Kong*, see Cynthia Erb, *Tracking King Kong: A Hollywood Icon in World Culture* (Detroit: Wayne State University Press, 1998), 121–54.

27. Makeba, *Makeba*, 68.

28. Todd Matshikiza, *Chocolates for My Wife* (London: Hodder & Stoughton, 1961), 112.

29. Kenneth M. Cameron, *Africa on Film: Beyond Black and White* (New York: Continuum, 1994), 113.

30. Benson Dyantyi, "Last Days of Sophiatown," *Drum*, November 1959, 43.

31. Modisane, *Blame Me on History*, 3.

32. Ibid., 5.

33. Ibid., 10.

34. *Drum*, January 1959, 27.

35. In his memoir, Modisane, who played the part of a gangster named Shark in Athol Fugard's *No Good Friday*, staged a few months prior to *King Kong*, states that he based his performance partly on Brando's work in *The Men*, indicating Brando's status as figure of admiration for both white and black South Africans. He also states that he studied Brando's technique because of a fascination with the Method, suggesting that for a black South African frequently barred from productions at white theatres, one place to learn Method acting was at the movies (*Blame Me on History*, 289–90).

36. Nathaniel Nakasa, "Criminals Without Crime!" *Drum*, April 1959, 23.

37. Makeba, *Makeba*, 71.

38. Modisane, *Blame Me on History*, 296.

39. Makeba eventually became a prominent antiapartheid activist on the world scene. In 1987 she appeared with Paul Simon during part of the "Graceland" tour.

40. Coplan, *In Township Tonight!*, 162.

41. Ibid., 176.

42. Glasser, *King Kong: A Venture*, 61.

43. Advertisement, *Drum*, May 1959, 3; "Miriam: Star of Venice," *Drum*, November 1959, 65.

44. Matshikiza, *Chocolates*, 121.

45. Ibid., 121.

46. Ibid., 110.

47. Gibson Kente, *Too Late*, in *South African People's Plays*, ed. Robert Mshengu Kavanagh (London: Heinemann, 1981), 113. For Kavanagh's discussion of Kente's work, see *Theatre and Cultural Struggle*, 113–44.

Chapter 9

Traveling Players:
Brazilians in the Rouen Entry of 1550

Claire Sponsler

I n October of 1550 the town of Rouen staged a festival in honor of Henri II and his wife Catherine de Médici on the occasion of their entry into Rouen. This festival was one of the most elaborate and spectacular in a performance genre known for its over-the-top theatricality and spare-no-expense extravagance.[1] The festival included such typical displays as pageants and *tableaux vivants* featuring Roman gods, muses, and nymphs and was marked by the commingled themes of flattery and persuasion typical of royal entries. According to the various contemporary accounts still extant, whose number and detail attest to the fascination the festival inspired, there were marvelous animals like unicorns and elephants side by side with battling gladiators, a mock sea fight between Portuguese and French warships, and a procession of captives won in recent battles. The festival's *pièce de résistance,* however, for modern scholars and apparently for sixteenth-century spectators as well, was a meticulously re-created "Brazilian" village, built at the Faubourg Saint-Sever on the banks of the Seine just outside the city walls.

Although the Rouen entry has attracted a good deal of scholarly attention, including from new historicists specifically interested in cultural negotiation and colonialist co-optation, no study has explored the dynamics of cross-cultural performance at work in the re-created Brazilian rain forest of the Rouen festival.[2] While it is not difficult to see how the drama that was staged within this village can be read as an act of cultural poaching, which stole not only the exotic New World setting but also Brazilians themselves, replanting the whole lot within the boundaries of Europe, that is only a part of the story. For, side by side with the fifty Brazilian villagers

168

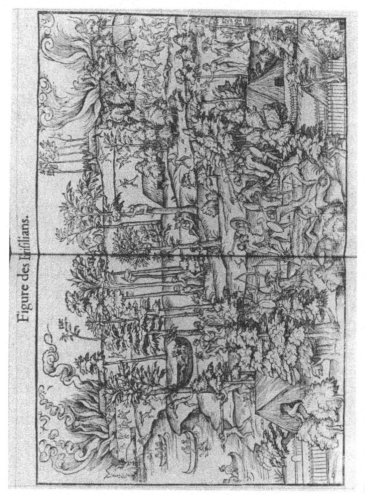

Figure 9.1. Figure des Brisilians. By permission of the Collection de La Bibliothèque municipale de Rouen (Photographies Didier Tragin/Catherine Lancien)

who were imported for the performance, there were some 250 Norman sailors, dressed up to look like Brazilians, who also played parts in this drama. If the language of the fullest extant account, *C'est la deduction,* can be taken as an accurate record, issues of imposture, verisimilitude, hybridity, and cultural exchange were at the forefront of the performance, complicating the apparent power relations and paths of domination in the performance.[3]

Stephen Greenblatt has argued that "[t]he native seized as a token and then displayed, sketched, painted, described, and embalmed is quite literally captured by and for European representation" and thus "caught up in a complex system of mimetic circulation."[4] While such a description allows for the possibility of reciprocal exchange—circulation, after all, is circular—in most analyses, the exchange tends to be viewed as one-sided, with all the power in the hands of those who do the seizing and representing. This model has been challenged in recent years by advocates of an alternate view of cultural interactions, one based less on capture and circulation—which are, from the side of the victim at least, passive acts—than on complicity and travel, both of which accord the "native" a role of at least limited action and choice. James Clifford has recently called attention to what he labels the "Squanto effect." Squanto, of course, was the Indian who greeted the Pilgrims at Plymouth in 1620 and who helped them survive the winter. He spoke English, a Patuxet just back from a stay in Europe. For Clifford, Squanto represents the hybrid "native"—a person who is oddly familiar but also different, precisely because of that unprocessed familiarity.[5] As Clifford accurately notes, the trope of the hybrid native has become increasingly common in accounts of postmodern cultural interactions; in these accounts, the roles of hybrid native, informant, and traveler overlap, and in so doing offer a different way of thinking about cultural contact. Additionally, these accounts "free" the "native" from the geographic and conceptual place in which he has tended to be fixed in Western discourses. As Arjun Appadurai has argued, the term "native" itself has imprisoned non-Western people through a process of representational essentializing that has confined natives to the places where they belong, even though groups untouched by contact with a larger world have probably never existed. Attributions of immobility have spawned a language of incarceration, Appadurai notes, which describes natives as trapped in places to which outsiders (explorers, colonizers, administrators, anthropologists, tourists) travel, but from which the native cannot escape.[6] Ethnography, as Clifford notes, has had a similar tendency, marginalizing a culture's external relations by looking nearly exclusively at "natives in their villages." The emergent model of culture-as-travel-relations described by Clifford has begun to make cultures look less like isolated villages and more like teeming hotels—transient, hybrid environments where cultural encounters routinely take place.[7] In a similar vein, the notions of *détour* and *retour* proposed by Edouard Glissant in his study of Martinique are productive for theories of a postcolonial habitus, since they describe the comings and goings of people alternately excluded and welcomed both at home and abroad.[8]

This model of culture-as-travel-relations, which sees the native as traveler, is particularly apt for the Rouen performance, at which not just performers but also spectators were transients, temporarily participating in the creation of an imagined and distinctly hybrid community. In the remainder of this essay, I would like first to describe the performance itself—the "plaisant Spectacle" (L1r) witnessed by Henri and Catherine—and then to consider it as a cross-cultural performance that participated in the construction of a hybrid community whose features, while geared toward encouraging specific economic and political undertakings, also had a wider resonance for European and New World relations.

Like the other interconnected parts of the entry, the Brazilian village was conceived by the group of provincial elites who funded and planned the entry, Rouen's town council. Planning for the entry began on June 12, 1550, with the town council coercing cooperation and financial support from the residents of Rouen. Attesting to the care that went into the planning, the councilors stipulated that all of the entertainments had to be approved by them and organized into a coherent whole; twelve masters of ceremony were appointed to oversee the event.[9] One of the men they chose to create the various spectacles was Claude Chappuys, writer, chamberlain, courtier, librarian to François I, and canon of the Cathedral de Notre Dame in Rouen from 1537 to 1575.[10] Chappuys had assisted at François I's funeral and was at his successor's coronation. Perhaps more to the point, Chappuys had apparently played a large part in organizing Henri's royal entry into Paris in 1549.

The first part of the entertainments showcased Rouen's civic order, with processions of clergy, merchants on horseback, crossbowmen, citizens, the trades, soldiers, and children. Pageants reenacting Henri's recent military exploits followed, including a procession of captives. And finally Henri himself rode in procession toward the Seine. There he mounted one of two scaffolds that had been erected and looked out on the re-created Brazilian village. Afterwards, the king and his company moved toward the bridge, where a large "rock" had been fabricated, within which Orpheus sat playing his harp accompanied by the Muses, and Hercules fought the Hydra (a placard explained that the king was another Hercules). As the king crossed the bridge, a sea battle was staged, representing the fight between Portugal and France for Brazil. Henri then passed under two elaborate arch structures, decorated with images and verses symbolizing France's connection with Troy, and came to the final pageant modeled on the Elysian fields. The combined force of these various spectacles was to praise and admonish Henri while also presenting Rouen in a favorable light as a prosperous, devout, and powerful town.

According to *C'est la deduction*, which was published one year after the performance and devotes two full pages of text and a double-page woodcut to the performance, the "Brazilian" village constructed outside Rouen covered a plot of land approximately 200 paces long by thirty-five wide, formerly a field of shrubbery and willows. A color illustration included in *C'est la deduction* shows the village lying to the left side of the bridge over the Seine, which Henri and Catherine would cross during the course of their entries after having watched the performance staged for them in the re-created rain forest. For the performance, boxwood and other shrubs had been planted in the field, as had fruit trees, to give the lush appearance of a Brazilian jungle. Tree trunks painted red and topped with branches and fronds had been placed among the local trees in imitation of a Brazilian forest; their branches were filled with parrots and monkeys, which had been imported by the bourgeois of Rouen from Brazil. Huts ("loges ou maisons," K4v) constructed of tree trunks and thatched with leafy boughs had been erected at each end of the open meadow, surrounded by palisades of sharpened stakes. Standing on scaffolding erected for his benefit, Henri was able to look on as within the village some 300 men and women went about their ordinary affairs—cutting wood, bartering with French sailors for axes, chisels, and fishhooks, hunting birds and monkeys, or swinging in their hammocks.

Most of the villagers—some 250 of them—were impersonated by Norman sailors who had been hired to play the part of savages. The other fifty were natives who had been brought from Brazil by a merchant from Rouen—"freschement aportez du pays" (K3v). They were probably members of the Tupinamba people, with whom the merchants of Rouen had established trading relations. All of the performers were, according to both text and illustration, naked—the text says they were "tous nudz" (K3v). The "true" Brazilians ("naturelz sauvages," K3v) had pierced ears, cheeks, and lips; they wore earrings of polished white and emerald stones. The Norman sailors were carefully made to appear to be Brazilians as well. They were said to have been chosen, according to *C'est la deduction*, because they had lived in Brazil and knew the native gestures and customs, so much so that they seemed as if they, too, were natives of the same place ("comme s'ilz fussent natifz du mesmes pays," K3v).

In this reconstructed rain forest (fig. 9.1), the "Brazilians" enacted a scene of industrious commerce, cutting down trees and carrying them to the river, where sailors negotiated a price and the timber was loaded onto ships that flew flags of the merchants of Rouen and banners of the black

and white fleur-de-lis of Henri. The language of *C'est la deduction* makes
clear that this scene was intended to portray the customary trading activities of the Normans in Brazil, stressing that the sailors behaved as they "ont
accoustume faire" (K4r) and that the transactions were carried out "selon
leur usage & maniere de faire" (K4r). As with the performers themselves,
the actions they engage in are carefully described in both text and woodcut as genuine, suggesting the high premium put on the creation of an accurate representation of real persons and events. This emphasis on
verisimilitude plays an important role in the vision of European-Brazilian
relations the performance attempted to construct, and, not incidentally, in
the model of cultural exchange it espouses, as I shall discuss shortly.

The idyllic and diligent scene was abruptly interrupted, *C'est la deduction* says, as a group of Indians called "Tabagerres," incited by their king,
"Morbicha," attacked a rival group, the "Toupinabaulx." Shooting arrows
and wielding clubs, the two tribes fought fiercely, until the "Toupinabaulx"
gained the upper hand, routed their attackers, and burned their hut to the
ground. *C'est la deduction* notes that according to those who had traveled
to Brazil, the staged battle was so realistic that it was a sure simulacrum of
the truth (a "certain simulachre de la verité," K4r). It was also pleasing to
Henri, as might have been expected given his taste for military exploits.[11]
Additionally, since this mock battle showed the Tupinamba (allies of the
French) triumphing over the Tobajaro (allies of the Portuguese), it would
have been doubly pleasing in its message of French dominance over a foe,
a pleasure intensified in the entertainment that immediately followed, featuring a naval battle between a French and a Portuguese ship.[12] This battle between "Brazilian" villagers was repeated on October 2 for Catherine
on her entrance, although *C'est la deduction* remarks only that the queen
and her entourage enjoyed the "esbatementz & schyomachie des sauvages
du bresil" (P3r), before crossing the bridge.

The double-page woodcut of this scene in *C'est la deduction* offers both
an ethnographic snapshot of the village with its inhabitants and a dramatic
record of the performance, thus suggesting both stasis and dynamism. Its
ethnographic gaze captures men and women in a series of what seem intended as typical scenes of everyday Brazilian life, and the spectator is invited to take up the voyeuristic and distanced position so often proffered
by visual representations of exotic others in the early modern period.[13]
But the woodcut also offers a dynamic reenactment of the performance itself, closely following the description of the performance in the text of
C'est la deduction, showing the cutting of timber and the trading activities
on the river, the battle between the two sets of villagers, and the burning
of the hut of the defeated Tobajaro. This simultaneous staging of the linear
events of the drama to some extent undercuts the image of voyeuristic

spectator gazing on captured specimens. Although the European spectators at the Rouen performance gaze at the spectacle laid out before them, they nonetheless do so within a set of theatrical conventions that encourage a linking of spectator and performer. It was this logic of implied connection that made the processions of the citizenry of Rouen that occurred earlier in the entry so potent: Henri was meant to register not just the wealth, power, and probity of the townsmen as they marched before him, but also his bond with them as their ruler. The dynamics of theatrical spectatorship in the sixteenth century assume a fluid border between observer and performer, reality and performance.[14]

Some scholars have argued that the "Brazilians" in this spectacle would have been understood symbolically, as a parody of ideals of humanist behavior or as types of a New World Hercules,[15] but there is good reason to believe that they might have been taken as precisely what *C'est la deduction* states that they were attempting to impersonate—Brazilian villagers engaged in a commercial exchange with Norman mariners. Although the entry took place five years before the unsuccessful attempts by the French to establish a colony in Brazil (1555–60), it can nonetheless be read in light of Franco-Brazilian relations at midcentury, as Jody Greene has persuasively done.[16] For at least fifty years before Henri's entry into Rouen, Normandy was involved in fishing and commercial ventures across the Atlantic, and Norman sailors had established trading relations with Tupi-speaking villagers on the Brazilian coast—where they traded primarily for brazilwood, but also parrots, monkeys, and jaguar skins—despite hostile competition from the Portuguese.[17] As Greene notes, until 1550 Norman merchants resisted efforts to colonize Brazil, since that would have required royal permission and the oversight and control (including taxation) that would have accompanied it, preferring instead to shoulder the risks of unprotected sea travel in order to reap the profits of unregulated trade.[18] Merchants from Normandy traded directly with Brazilians throughout the sixteenth century, but without royal protection, since Brazil officially belonged to Portugal; hence, their trading was technically illegal, as Greene observes.[19]

In the late 1540s, Henri revoked various privileges the Norman merchants had previously enjoyed, thus threatening the profitability of voyages to Brazil and providing the Rouen merchants with a reason for mounting a special plea for their cause. The Rouen performance, in Greene's view, re-created not Brazilian culture, but "a carefully composed representation of the harmonious trading relations between the French and their Tupinamba associates,"[20] a representation designed to assure that the favorable trading ties the merchants of Rouen had with Brazil were maintained, despite Henri's recent ascent to the throne and all the potential disruption that might forebode. The organizers of the entry placed Brazil at the center, Greene argues, since they sought to pressure the king into supporting the continuation of their trading relationship with the New World, more specifically, with the people living on the Brazilian coast between Guanabara Bay and Pôrto Seguro. The efficacy of the "Brazilian" spectacle is perhaps indicated by the fact that the year after the Rouen entry, Henri repealed his Brazilian trading regulations, although for the first time imposing taxation on brazilwood.[21]

Within this context of mercantile coercion, what might the role of the performers—both Brazilian and non-Brazilian—have been? For one thing, the "Brazilians" in the spectacle might have seemed less strange and exotic than we might at first glance assume. C'est la deduction does not suggest that the "sauvages" were unusually strange or in any way monstrously incomprehensible, although it does refer to such exoticizing features as their nudity and their use of bodily adornments. Most of the description in C'est la deduction, however, matter-of-factly relates how the villagers cut timber and trade it with the Normans, and then, albeit with heightened emotion, recounts the battle scene. The account's frequent repetition of claims that the representation was so accurate as to be practically real implies both a familiarity with Brazilian customs that could serve as a ground for judging the accuracy of this particular portrayal and a privileging of representational realism. Similarly, while the woodcut shows allegiance to both pastoralism and the "wildman" tradition, it, too, in its panoramic inclusiveness and detailed representation of a variety of simultaneously depicted activities—hunting, dancing, lounging in a hammock, strolling through the forest, hauling timber, and fighting—conveys an overall sense of ordinariness and verisimilitude, a sense underscored by the written account.

Familiarity might have been acquired not just via accounts brought back by Norman sailors, but by acquaintance with actual Brazilians, a number of whom found their way, whether voluntarily or not, to Normandy in the first half of the sixteenth century. The first encounter of Europeans with Brazilians was a voyage by some thirteen ships under the command of Pero Alvares Cabral, who left Lisbon on March 8, 1500. The fleet sailed along the coast for nine days, then left, but recorded the visit of two young Brazilian men to Cabral's flagship.[22] As Green notes, the encounter "introduces such stock elements of the literature of early encounters as miscommunication, mimicry, and exaggeration by the indigenous population designed to impress the Europeans."[23] Caminha reports that one of the young men, along with other Brazilians, asked to accompany the Portuguese; Caminha says that "if we had invited them all, they would all have come"; in fact, the Portuguese take only four or five.[24] No more is known about these visitors, if they even made it to Portugal, but they at least offer evidence (if Caminha is to be believed) of Native American desires to, in Greene's apt phrase, "turn the European observer into a tourist attraction."[25]

The practice of bringing back Native Americans was widespread.[26] They were brought back by Columbus, Cabral, Vespucci, and Gaspar Corte-Real in 1501; by Thomas Aubert in 1508; by Cortés in 1528; and by Jacques Cartier in 1534 and 1536.[27] The captives Cortés brought back with him performed acrobatic tricks for Pope Clement VII in Rome.[28] Aubert, from Dieppe, brought seven men to Rouen from what was probably Newfoundland. In 1528, Cartier's wife stood as godmother to a child of a woman named "Catherine de Brezil" in Saint-Malo, though why Catherine was brought to Europe is unknown.[29] When the Norman merchant Jean Ango arranged the entry of François I into Dieppe in 1534, he had Brazilians in his house who greeted the king.[30]

How willingly did these people arrive on European shores and perform for or interact with Europeans? In all likelihood, few must have come voluntarily (although see once again the example of the Brazilians brought back by Cabral). The motives for capture were numerous, including the need to bring home proof, the desire to spur the curiosity of future investors, and demand for interpreters, all of which would have made seizure of natives nearly irresistible.[31] Despite these reasons arguing for the predominance of involuntary captivity, at least some captives must have been voluntary travelers; to deny this possibility, as Greene aptly notes, is to deny all chance for agency on the part of those from the invaded cultures. Moreover, disallowing voluntary travel misrepresents the complexity of cultural contacts, in which individuals and groups, even as early as the sixteenth century, participated in comings and goings that blurred notions both of home and abroad and of captive and traveler.

Americans who came into contact with Europeans might have had their own reasons for wishing to accompany them home, including the quest for technological information (about the use of iron or gunpowder), which Europeans were reluctant to part with in a colonial setting, or trading advantages (including the advantage of learning European languages). Curiosity, which Greene reminds us could have been a two-way street,[32] might also have provided a motive for journeys to Europe. The best-documented example of a Brazilian who apparently willingly traveled to Europe is that of the fifteen-year-old "prince" Essomeric, who, along with an older companion named Namoa, returned to Normandy with Binot Paulmier de Gonneville in 1505.[33] In June 1503, de Gonneville set sail from the Norman port of Honfleur in the *Espoir* with sixty men, in search of a route to Asia; blown off course, he landed on the coast of Brazil (which he took to be Australia), where he spent the next six months trading with Carijó villagers. According to the deposition given by crew members on their return, the headman of the village wished to have his son accompany the ship on its return voyage, being assured that the youth would be brought back within two years. After surviving storms, pirates, and sickness (which took Namoa's life), the ship finally reached Honfleur in 1505. Essomeric remained in Europe for the rest of his life and was essentially adopted by the heirless de Gonneville, who gave him his surname and part of his lands, and arranged his marriage to a rich family member.

The inhabitants of the New World were also at times required to play themselves—or fantasized versions of themselves—on European stages, often to serve the larger political agendas being enacted. Examples include Henry VIII's wildman characters in the Twelfth Night celebrations of 1515; the royal entry of Eleanor of Toledo into Florence as the future duchess in 1539, when she passed under an arch depicting Charles V as conqueror of Mexico and Africa; and the tournaments of 1584 for Elizabeth I, which included servants dressed like savages and Irishmen. Such performances allowed monarchs to display and ratify political control through manipulation of cultural codes of magnificence using the figure of the vanquished savage as a central actor.[34] These performances might well have expressed a fascination with difference, but we should not ignore the possibility that they could also represent an opportunity to put on an alien identity, an opportunity that had repercussions for existing power structures and patterns of social relations. In the case of the Rouen entry, in which Europeans imitated "sauvages" and then performed side by side with them, these two dynamics probably coexisted, with the performance allowing at least some spectators to enjoy the pleasures of voyeuristic viewing of exotic people, while letting other spectators as well as the European performers have the chance to construct imagined lines of identification and familiarity between themselves and the "sauvages."

As I have argued elsewhere, even within a New World context, performance played a more important part than has usually been assumed in mediating initial contact between Europeans and Americans.[35] Recent studies by theatre historians have begun to delineate how extensive performance was in the conquest of Spanish and French America.[36] Building on Aztec ritual ceremonial, the Franciscan missionaries who arrived in what is now Mexico in the spring of 1524—and later Dominicans and Jesuits—sponsored a didactic religious theatre that used indigenous actors performing in the *lingua franca* of Nahuátl for an indigenous audience. The earliest recorded Spanish-American performance was a conversion play based on the life of Saint Paul, performed in 1530 in Nahuátl by an Aztec cast in Mexico City. Biblical and saints' plays, historical dramas, and moralities followed. Drama also played a role in the French colonization of America. *The Jesuit Relations,* which chronicle the activities of French missionaries in the Great Lakes and St. Lawrence regions, offer scattered evidence of a variety of dramatic activity, including religious processions for the Assumption and Corpus Christi, processions in honor of royalty, passion plays, secular plays like Corneille's *Le Cid,* dancing, ballets, and feasts. The *Relations* give a particularly full description of a Three Kings pageant enacted in 1679 at the Huron Mission of Tionontate at St. Ignace de Missilimakinac. In this culturally complicated French-Huron collaboration, Hurons under Jesuit direction constructed a "grotto" inside the mission church for the "representation of the mystery" of the Nativity. At the climax of the performance, the three chiefs playing the roles of the three kings made public proclamation of their submission and obedience.[37] Carefully staged rituals of greeting, exchange, battle, and sovereignty formed the constituent features of this theatre of conquest. Colonizers acted roles, pretending to be the saviors and inheritors they imagined themselves by divine and royal right to be. Using a telling phrase, Thomas Hariot described himself in his *Briefe and True Report of the New Found Land of Virginia* (1588) as "an actor in the colony," suggesting his awareness of the role he had signed on to play as a member of the first British colony at Roanoke in 1585.[38]

European colonizers were also fascinated by native performances, which are mentioned frequently in early travel accounts. These performances tended either to be seen as strange, exotic, and grotesque—that is, as signs of otherness—or to be assimilated to European performances, thus bridging the gap between self and other. For example, James Rosier describes "the ceremonies of their idolatry," emphasizing the strange gestures, loud cries, grimaces, shrieks, and other demonic activities of the performers; John Gage likens a native dance to "our" morris dancing; and Theodore de Bry depicts native dancers so that they seem to be dancing around maypoles.[39] Jacques Le Moyne's description in his *Brevis Narratio* (1591) of the first encounter of René de Laudonnière with the Timucuan chief Satouriwa in Florida neatly captures this dual perspective, beginning by likening Satouriwa's train to a royal entry but then shifting gears to stress the animal nature of the chief.[40] As this last example suggests, part of the drive behind assimilating native performances to European ones was not only to deflate their alterity but also to reshape threats of aggression via the metaphor of play and theatricality. These performances, and others like them, hint at how important theatricality was to European-American encounters, offering a medium for negotiating some of the most charged moments of cultural contact within what Mary Louise Pratt has called the "contact zone," that space of colonial encounters where "peoples geographically and historically separate come into contact with each other and establish ongoing relations, usually involving conditions of coercion, radical inequality, and intractable conflict."[41]

As Greene observes, coerced though they might well have been, it is unlikely the fifty Brazilians in the Rouen entry were captives, since that would hardly have enhanced Norman trading relations there.[42] Moreover, *C'est la deduction* takes pains to stress the almost seamless way in which real and imitation Brazilians blend together in the performance, all playing their parts in essentially inseparable fashion. After initially establishing that the majority of the performers are Norman, the text then goes on to erase any difference between Normans and Brazilians within the performance context—all become "Tabagerres" and "Toupinabaulx"—and the erasure of difference is echoed in the woodcut as well. This indistinguishability suggests perhaps that all of the performers should be seen as precisely that—performers playing scripted roles. To see the Rouen performers in this light is to imagine that cultural exchange has been thoroughly achieved, however temporarily, within the performance space.

In European culture the role of the performer is one of the few available to outsiders, in part because it does not challenge stereotypes.[43] New World residents were usually expected to be not just the translators in colonial encounters, but also the performers of cultural exchange—witness Cortés' captives performing for Clement VII. The Rouen performance can be seen as an elaborate and literal "translation" effected collaboratively by Normans and Brazilians. An entire village is translated from Brazil to France, with the performers acting as mediating agents—some European, some Brazilian—whose characters and actions represent simultaneous images of exotic otherness and familiar contact. The Brazilians, and the Europeans impersonating them, thus become both self and other, participating in a staged event that oscillates between defamiliarization (making the familiar European countryside exotic and foreign) and assimilation (drafting Normans and Brazilians into a performance geared toward strengthening both European and Brazilian material interests). The performance can in this way be seen as exploiting exotic otherness and the dialectics of difference and sameness within a specific context and for immediate local purposes, although we need not assume that the performance's potential meanings would be limited just to that local context. Indeed, the fact that the Rouen entry quickly jumped media and was reproduced in both written and visual texts that could be readily disseminated to a wider public indicates precisely how efficiently an ephemeral, provincial performance could become caught up in wider circuits of representation and hence signification.[44]

The Rouen performance falls into a distinct category of cultural encounter, what Urs Bitterli has called the "direct encounter," in which Europeans encountered non-Europeans under circumstances that forced a relatively unmediated form of cultural interaction.[45] In the direct encounter, there exists the possibility for "going native" and experiencing the integration of self with other. Although the direct encounter was for the most part an infrequent form of contact between Europeans and non-Europeans in the early years of colonization, traces of it can be seen in various accounts, particularly in captivity narratives by Europeans. The Rouen spectacle can be seen as a species of direct encounter in at least two ways. First of all, from the point of view of the 300 performers, the spectacle not only allows but insists on unmediated, direct contact between Normans and Brazilians. As *C'est la deduction* implicitly recognizes, the success of the performance hinges on how effectively Normans and Brazilians can become one, joining together in a collaborative performance as "Tabagerres" and "Toupinabaulx." It is, for instance, striking that the parts are not divided up along antagonistic lines, with "true" Brazilians playing one set of villagers and Normans taking on the opposing role. Instead, Normans and Brazilians apparently played side by side, impersonating both sets of villagers. In effect, then, no one is the "true" Brazilian, since all are playing adopted parts, a process that must inevitably have lessened differences, at least for as long as the performance lasted, between European and Brazilian.

The numerous instances of cross-cultural impersonation in theatrical performances of the fourteenth through sixteenth centuries, whether in the cycle plays (in such characters as Jews, the magi, and Herod) or in other entertainments such as royal entries, civic processions, and court disguisings, suggest that the Rouen performance was not unique in this regard. In a 1501 performance in France, for instance, a player dressed as a Turk was refused a dance and at the Vow of the Pheasant in 1454, a Turk/Infidel was shown holding Saint-Eglise prisoner.[46] On the occasion of the wedding festivities for Prince Arthur in 1501, Sir Nicholas Vaux was "disguised" as a Turk or Saracen.[47] In 1510, the first year of Henry VIII's reign, there was a disguising in which Turks and "moryons" mingled with Russians and Prussians.[48] Edward VI performed as a Moor in a Shrovetide masque of Moors of 1548, and the Christmas festivities of 1550 at the English court included masques of Irishmen and Moors.[49] The King of the Moors and his morians were traditional figures in the London Lord Mayor's Show.[50] Two men disguised as Moors, with blackened faces, rich costumes, and white cloths around their heads as if they were Saracens or Turks, presented gifts at Isabella's 1389 royal entry into Paris, and citizens of London rode to Kensington to perform a mumming for Richard II in 1377 disguised as African or Eastern ambassadors, among other things.[51]

And, performing a somewhat different sort of drama in 1451 after Cade's rebellion, blackfaced "servants of the Queen of the Fairies" broke into the duke of Buckingham's park in Kent and took his bucks and does.[52] What is unusual about the Rouen performance, then, is not its use of impersonation of racial others, but rather the leap its impersonation took out of Europe and across the Atlantic. Certainly the fairly long-standing tradition of impersonation of racial others must have provided a horizon of expectations within which the "true" and imitation Brazilians of Rouen could have been contextualized.

From the perspective of its spectators as well, the Rouen spectacle can also be seen as a form of direct encounter. Exoticism is a mode of representation in which the culture in power transforms the other culture into an unreal and therefore unthreatening culture. But for the Normans in Rouen, the Brazilians posed no threat. There was no need to exoticize people with whom stable trading relations had been established and who had for twenty years been visiting, willingly or not, Norman towns. For the people of Rouen, Brazil was real, and Brazilians neither strange nor stereotypical (despite being styled "sauvages"). And it was precisely this familiarity that allowed the town council to devise a collaborative performance for the entertainment and enlightenment of the real outsiders—Henri and his entourage. Although Henri watched the scenes in the "Brazilian" forest from the distanced vantage point of scaffolding, he viewed this entertainment as part of a larger and intensely familiar whole, well known to Henri thanks to the more than thirty entries he participated in during the years following his coronation.[53] While some level of exoticism might have informed Henri's viewing of the "Brazilian" spectacle, the containment of that spectacle within the conventional totality of the whole entry must have worked to familiarize the spectacle. Additionally, as I mentioned earlier, the model of engaged spectatorship still prevalent in the sixteenth century encouraged a kind of direct contact between audience and performers that worked to break down the distance required for a genuinely exoticized representation.

By asking insiders to temporarily become outsiders by imitating the objectified other, making the object a subject, and inhabiting that subject, the Rouen entry explored the dynamics of exclusion and domination, of sameness and difference. As Homi Bhabha has remarked, "Mimicry is at once resemblance and menace."[54] The mimicry of a savage other—an other who was also, strangely, a trading partner and fellow conspirator in the project of enlisting Henri II's support—allowed the illusion of that menacing other's containment, even though such containment might only have been a refiguration of the conflicts and anxieties inherent in colonial and imperial relations. But the primary function of the "sauvages" in the

Rouen entry, particularly in the minds of the citizens of Rouen, was, finally, to contribute to the construction of a fictive cultural coherence; this spectacle thus contributes to an emerging ideology of colonial relations in sixteenth-century Europe, one marked by highly ambivalent desire and appropriation of the other.

As a performance, then, the Rouen entry seems thoroughly and undeniably cross-cultural, both in terms of interactions among European and non-European performers and in terms of the dynamics of spectatorship, which in this case sought to encourage direct and sympathetic contact between the French king Henri and his Norman subjects. For both performers and spectators at the Rouen entry, hybridity must have been a lived state of being, with the citizens of Rouen aware of themselves as attached both to France, from which they nonetheless remained separate, and to Brazil. The performance they offered Henri II in 1550, a performance at which Henri and his entourage may have been as much outsiders and transients as the fifty Tupinamba performers, demonstrates how a community—one linking the mutual interests of French king, Norman townspeople, and Brazilian villagers—could be imagined through a cross-cultural performance. Although it would be patently false to say that the interests of the Tupinamba dominated, it seems equally wrong to suggest that their interests played no part or that they were treated simply as exotic others displayed for the pleasure of the European gaze. Instead, what the Rouen entry suggests is that a model of cultural contact that sees natives as travelers and cultures as hotels meshes better with the realities of early cross-cultural performances than a model based on unilateral cultural co-optation. In Rouen in 1550, Brazilians played their parts *alongside,* not *for* Normans, in what stands as one of the earliest performances bringing together transatlantic cultures in a collaborative, cross-cultural theatrical effort.

Notes

1. For general discussions of the royal entry as a theatrical and political event, see Lawrence Bryant, *The King and the City in the Parisian Royal Entry Ceremony: Politics, Ritual, and Art in the Renaissance* (Geneva: Droz, 1986); Gordon Kipling, *Enter the King: Theatre, Liturgy, and Ritual in the Medieval Civic Triumph* (New York: Clarendon, 1998); and Roy Strong, *Art and Power: Renaissance Festivals, 1450–1650* (Suffolk: Boydell Press, 1984). *Les Entrées royales françaises de 1328 à 1515,* ed. Bernard Guenée and Françoise Lehoux (Paris: Éditions du Centre National de la Recherche Scientifique, 1968), offers an overview of French royal entries in the late medieval and early modern period.

2. The Rouen entry has been discussed by, among others, Josèphe Chartrou, *Les Entrées solennelles et triomphales à la renaissance, 1484–1551* (Paris: Presses Universitaires de France, 1928); Ivan Cloulas, *Henri II* (Paris: Boydell,

1984), 274–94; Victor E. Graham, "The Entry of Henri II into Rouen in 1550: A Petrarchan Triumph," *Petrarch's Triumphs, Allegory and Spectacle*, ed. Konrad Eisenbichler and Amilcare Iannucci (Ottawa: Dovehouse, 1990), 403–13; and Margaret M. McGowan, "Forms and Themes in Henri II's Entry into Rouen," *Renaissance Drama* 1 (1968): 199–252.

3. The sixteenth-century accounts of Henri's entry include *C'est la deduction du Somptueux ordre, plaisantz spectacles et magnifiques theatres dresses et exhibes, par les citoiens de Rouen* . . . (Rouen: Robert et Jehan dictz Dugord, 1551), which contains woodcuts of the entertainments and is available in a facsimile edition introduced by Margaret McGowan; see McGowan, ed., *L'Entrée de Henri II à Rouen 1550* (Amsterdam: Theatrum Orbis Terrarum, 1977). All quotations are from this account. Another account is *L'Entrée du Roy nostre sire faict en sa ville de Rouen* . . . (Rouen: Robert Masselin, 1550), which has been reproduced by A. Beaucousin (Rouen: Société des Bibliophiles Normands, 1882). A third account is *L'Entrée du très Magnanime très Puissant et victorieux Roy de France Henry deuxism de ce nom en sa noble cité de Rouen* . . . (Bibliothèque Municipale de Rouen, Ms.Y.28), available in a nineteenth-century edition by S. Merval (Rouen: Société des Bibliophiles Normands, 1868). A fourth account can be found in the deliberations of the Hôtel de Ville (Bibliothèque Municipale de Rouen, Registre A.16, délibérations, fols. 110–15). There is also a compilation of Bibliothèque Municipale de Rouen, Ms.Y.28, and the woodcuts from *C'est la deduction* called *Les Poutres et figures du sumptueux ordre plaisantz spectacles, et magnifiques theatres dressée et exhibés par les citoiens de Rouen* . . . *Faictz à l'entrée de la sacree Maiesté du trés chretien Roy de France, Henry second* . . . (Rouen: Jean Dugort, 1557).

4. Stephen Greenblatt, *Marvellous Possessions: The Wonder of the New World* (Chicago: University of Chicago Press, 1991), 119.

5. James Clifford, *Routes: Travel and Translation in the Late Twentieth Century* (Cambridge: Harvard University Press, 1997), 28.

6. Arjun Appadurai, "Putting Hierarchy in Its Place," *Cultural Anthropology* 3 (1988): 36–37.

7. Clifford, *Routes*, 26–30.

8. Edouard Glissant, *Le Discours Antillais* (Paris: Seuil, 1981), 28–36.

9. See McGowan, ed., *L'Entrée*, 13.

10. Archives Municipales de Rouen, Reg. A.16, fol. 78. See the detailed discussion of Chappuys' influence on the Rouen entry by Michael Wintroub, "Civilizing the Savage and Making a King: The Royal Entry Festival of Henri II (Rouen, 1550)," *Sixteenth Century Journal* 29 (1998): 465–94, esp. 482–83. Wintroub analyzes the entry, and particularly its use of the figure of Hercules, as a mechanism for "civilizing" Henri by modeling for him (humanist) alternatives to the feudal values of a warrior king.

11. *C'est la deduction*, L1v, describes Henri's pleasure at the Brazilians' mock battle. For his interest in warfare, see Frederic Baumgartner, *Henry II: King of France, 1547–1559* (Durham: Duke University Press, 1988), 40.

12. *C'est la deduction*, M2r. See also the account by the English ambassador describing this sea battle, which was understood to enact the fight between Portuguese and French for Brazil.

13. Steven Mullaney, *The Place of the Stage: License, Play, and Power in Renaissance England* (Chicago: University of Chicago Press, 1988), 67–68, asserts that the Rouen performance performed this sort of exoticizing function for its European spectators, including Henri and Catherine, whom Mullaney argues were invited to see themselves in the figures of the couple wearing crowns and lying in a hammock in the center of the woodcut.

14. For a discussion of late medieval theatrical spectatorship, see Claire Sponsler, "The Culture of the Spectator: Conformity and Resistance in Medieval Drama," *Theatre Journal* 44 (1992): 15–29.

15. Lawrence Bryant, *Politics*, 150, has argued that the Brazilian king is a parody of the ideal of the humanist orator king, since the Brazilian king fails to lead his people through eloquence and knowledge, instead inciting them to warfare. Wintroub, "Civilizing the Savage," 494, argues that the Brazilians in the Rouen entry were seen as types of a New World Hercules—as was the Brazilian king, Quoniambec, an "utterly alien being—a cannibal, a giant, a savage whose body was covered with tattoos and whose face was pierced and set with polished stones of emerald and green—was thus also a Hercules, and as such, a mirror, despite his 'otherness', of kingship and nobility."

16. Jody Greene, "New Historicism and Its New World Discoveries," *Yale Journal of Criticism* 4 (1991): 163–98.

17. For a history of French involvement in the New World, see Bradford E. Burns, *A Documentary History of Brazil* (New York: Knopf, 1966); John Hemming, *Red Gold: The Conquest of the Brazilian Indians, 1500–1760* (Cambridge: Harvard University Press, 1978) and Charles-André Julien, *Les Voyages de découvertes et les premiers établissements, XVè-XVIè siècles* (Paris: Presses Universitaires de France, 1948). The two sixteenth-century accounts of the French in Brazil are Jean de Léry, *Histoire d'un Voyage fait en la Terre du Brésil* (La Rochelle, 1578), and André Thevet, *Les Singularitez de la France Antarctique, autrement nommée Amérique: & de Plusieurs Terres & Isles Découvertes de Nostre Temps* (Paris, 1558).

18. Greene, "New Historicism," 180.

19. Ibid., 173.

20. Ibid., 170.

21. Ibid., 170, 187.

22. The encounter is described in a letter to King Manoel I by the chronicler Pero Vaz de Caminha; see Burns, *Documentary History*, 20–26.

23. Green, "New Historicism," 173.

24. Burns, *Documentary History*, 25.

25. Greene, "New Historicism," 174.

26. For the outlines of this practice, see Christian Feest, *Indians and Europe: An Interdisciplinary Collection of Essays* (Aachen: Rader Verlag, 1987); and Car-

olyn Thomas Foreman, *Indians Abroad, 1493–1938* (Norman: University of Oklahoma Press, 1943).

27. See Samuel Eliot Morison, *The European Discovery of America,* 2 vols. (New York: Oxford University Press, 1971–74), for a discussion of these and other cases.

28. See Bernal Díaz del Castillo, *The True History of the Conquest of Mexico,* trans. Maurice Keatinge (London: Harrap, 1927), 2:498–504.

29. For Aubert, see Eusebius, *Chronicon* (Paris: 1512); for Cartier, see F. Joüon des Longrais, *Jacques Cartier: Documents Nouveaux* (Paris, 1888), 15–16. Both instances are noted in Greene, "New Historicism," 175.

30. Paul Gaffarel, *Histoire du Brèsil Français au seizième siècle* (Paris, 1878), 83; cited in Greene, "New Historicism," 195, n49.

31. Feest, *Indians and Europe,* 614.

32. Greene, "New Historicism," 175–76.

33. See *Binot Paulmier de Gonneville. Campagne du Navire l'ESPOIR de Honfleur 1503–1505,* ed. M. d'Avezac (Paris: Challamel, 1869), and the discussion in Green, "New Historicism," 176–77.

34. See the perceptive discussion of these performances in David Richards, *Masks of Difference: Cultural Representations in Literature, Anthropology and Art* (Cambridge: Cambridge University Press, 1994), esp. 37–38.

35. See Claire Sponsler, "Medieval America: Drama and Community in the English Colonies, 1580–1610," *Journal of Medieval and Early Modern Studies* 28 (1998): 453–78.

36. The most important studies of Spanish colonial theatre are Othón Arróniz, *Teatro de evangelización en Nueva España* (Mexico: Universidad Nacional Autónoma de México, 1979); Max Harris, *The Dialogical Theatre: Dramatizations of the Conquest of Mexico and the Question of the Other* (New York: St. Martin's, 1993); Robert Potter, "Abraham and Human Sacrifice: The Exfoliation of Medieval Drama in Aztec Mexico," *New Theatre Quarterly* 8 (1986): 306–12; and Marilyn Ekdahl Ravicz, *Early Colonial Religious Drama in Mexico* (Washington, D.C.: Catholic University of America Press, 1970), which also provides English translations of a number of the plays.

37. For a description of the performance, see *The Jesuit Relations and Allied Documents,* ed. Reuben Gold Thwaites, 73 vols. (Cleveland: Burrows Brothers, 1899–1901), 61:114–19. The Three-Kings performance has been discussed by Martin W. Walsh, "Christmastide Performance in Native New France," paper delivered at the International Society for Medieval Theatre (SITM) conference in Toronto, August 1995.

38. Thomas Hariot, *A Briefe and True Report of the New Found Land of Virginia* (London, 1588).

39. James Rosier, *A True Relation of the Most Prosperous Voyage Made this Present Yeere 1605 by Captaine George Waymouth in the Discouery of the Land of Virginia* (London: George Bisop, 1605), sig. C2v. De Bry's illustrations accompany the 1590 edition of Hariot's *Briefe and True Report,* published in Frankfurt. Also see Hariot's description of ceremonial tobacco use "all done with strange

gestures, stamping, somtime dauncing, clapping of hands, holding vp of hands, & staring vp into the heaues, vttering therewithal and chattering strange words & noises" (*Briefe and True Report*, sig. C3v).

40. *The Work of Jacques Le Moyne de Morgues*, ed. Paul Hulton, 2 vols. (London: British Museum, 1977), 120.

41. Mary Louise Pratt, *Imperial Eyes: Travel Writing and Transculturation* (New York: Routledge, 1992), 6. For a recent analysis of the role of writing, and more specifically print, as an integral part of early English voyages of discovery and conquest, see Mary C. Fuller, *Voyages in Print: English Travel to America, 1576–1624* (Cambridge: Cambridge University Press, 1995).

42. Greene, "New Historicism," 183.

43. Roger Abrahams, *The Man-of-Words in the West Indies: Performance and the Emergence of Creole Culture* (Baltimore: Johns Hopkins University Press, 1983), 48.

44. For a related discussion of how exotic otherness could slip the bounds of a purely local context, see Claire Sponsler and Robert L. A. Clark, "Othered Bodies: Racial Cross-Dressing in the *Mistère de la Sainte Hostie* and the Croxton *Play of the Sacrament*," *Journal of Medieval and Early Modern Studies* 29 (1999): 61–87.

45. Urs Bitterli, *Cultures in Conflict: Encounters Between European and Non-European Cultures, 1492–1800*, trans. Ritchie Robertson (Stanford: Stanford University Press, 1989), 3.

46. See Jean d'Auton, *Chroniques de Louis II*, ed. R. de Maulde la Clavière (Paris, 1889–95), 2:100–1, and Clarence D. Rouillard, *The Turk in French History, Thought, and Literature (1520–1660)* (Paris: Boivin, 1941), 22–24. Rouillard notes the contrast with Turks in the 1468 Bruges tournament, where instead of representing infidels, they added an exotic, spectacular element.

47. *The Receyt of the Ladie Kateryne*, ed. Gordon Kipling, EETS os 296 (Oxford: Oxford University Press, 1990), n. 544.

48. See Edward Hall, *Hall's Chronicle* (London, 1809), 513–14.

49. Sydney Anglo, *Spectacle, Pageantry, and Early Tudor Policy* (Oxford: Clarendon, 1969), 301.

50. For entertainments by the livery companies of London, see *A Calendar of the Dramatic Records in the Books of the Livery Companies of London, 1485–1640*, ed. Jean Robertson and Donald Gordon, Malone Society Collections 3 (London: Oxford University Press, 1954).

51. Froissart, *Chronicles*, trans. Geoffrey Brereton (New York: Penguin, 1968), 359, describes the "Moors" in the 1389 entry; see Glynne Wickham, *Early English Stages, 1300–1660*, 3 vols. in 4 (London: Routledge and Kegan Paul, 1981), 3:48–49, for a discussion of the 1377 mumming for the boy Richard II.

52. In addition to these performances, see also the many illustrations of black magi and others depicted with dark skin in Ruth Mellinkoff, *Outcasts: Signs of Otherness in Northern European Art of the Late Middle Ages* (Berkeley

186   Claire Sponsler

and Los Angeles: University of California Press, 1993). See also the illumi-
nation in the *Très Riches Heures* of Jean, Duc de Berry (New York: George
Braziller, 1969), 24, of David's vision of two prophets preaching to the people of the world, who are represented by Africans seated on the left, Caucasians on the right. The Africans have dark skin and earrings and are
dressed differently from the Caucasians.

53. See Denise Gluck, "Les Entrées provinciales de Henry II," *L'information
d'histoire de l'art* 10 (1965): 191–218.

54. Homi K. Bhabha, *The Location of Culture* (London: Routledge, 1994), 86.

Chapter 10

South of North:
Carmen and French Nationalisms

Robert L. A. Clark

I n a genre in which violent passion, sexual transgression, and the victimization of women tend to be the rule, Georges Bizet's *Carmen* still manages to be exceptional a century and a quarter after its controversial premiere in Paris at the Opéra-Comique on March 3, 1875. Few would have expected, and perhaps least of all Bizet, who died exactly three months later, before his masterpiece had achieved wide acclaim, that *Carmen* would practically become synonymous with French opera and occupy the unshakable place in the canon that it has enjoyed now for over a century. *Carmen's* lack of success at its premiere (or "failure," according to many) cannot be attributed to any single factor, but certainly two in particular may be singled out: the newness of the music, which many critics qualified as "Wagnerian" (a term perhaps best understood as indicating Bizet's defiance of the traditions of the *opéra comique*, which was considered a French national genre); and the "scabrous" subject, drawn from one of the best-known works of Prosper Mérimée. Bizet's heroine, although a rather tempered version of her counterpart in the Mérimée novella, is a study in transgression. Indeed, her every dimension is placed under the sign of the other: female; doubly foreign and racially other, as a gypsy in Andalusia; working class (she works in the tobacco factory at the beginning of the opera); sexually dissident in relation to the bourgeois mores of the day; and, finally, an outlaw (she and her friends run contraband). And so she remains, defiantly, to the end.

Being a gypsy is hardly the same, of course, as being a factory worker. The combination is actually something of a conundrum that is resolved on the level of plot when Carmen loses her position in the factory after a

brawl involving one of her coworkers. On the ideological level, despite the
different historical contingencies of these various subject positions, they
are collapsed into a single position of otherness by virtue of their opposi-
tionality to the values of northern European bourgeois culture. Similarly,
"east of west" and "south of north" participate in the same nineteenth-
century Orientalist construct. As Victor Hugo famously remarked in the
original preface to *Les Orientales*, his thoughts and reveries were, by turns,
"hébraïques, turques, grecques, persanes, arabes, espagnoles même, car l'Espagne c'est encore l'Orient; l'Espagne est à demi africaine, l'Afrique est a
demi asiatique" ["Hebraic, Turkish, Greek, Persian, Arabic, even Spanish,
for Spain is still the Orient; Spain is half African, Africa is half Asian"].[1] My
goal in this essay will be to elucidate this othering of *Carmen,* from
Mérimée's initial concept to its transformation in the opera by Bizet and
his librettists Henri Meilhac and Ludovic Halévy. But beyond the role of
Carmen's male creators, I will look to the women who brought this remarkable character to life on stage. How did Célestine Galli-Marié, Emma
Calvé, and their successors negotiate the role with which many of them
ultimately came to be identified? Finally, I will address the shifts in cultural
politics surrounding some of the more noteworthy productions of *Carmen,*
from the post-Commune anxieties around the premiere to the opera's triumphant entry into the repertoire of the Paris Opéra at the height of the
Cold War, in the early years of the Fifth Republic.

A Rake's Progress

Prosper Mérimée (1803–70) did not live to see the stage debut of his most
famous character, but it is no exaggeration to say that his ghost haunted
the wings of the Opéra-Comique in March of 1875.[2] The librettists' attempts to make the story and, especially, the title role more palatable to the
conservative bourgeois public that was the mainstay of the Opéra-
Comique were largely compromised by Célestine Galli-Marié's interpretation of the role, inspired by her reading of Mérimée's novella. Critics, in
turn, were quick to compare the libretto, most often unfavorably, with its
source, universally recognized as a "masterpiece," and the opera's heroine
with Mérimée's original conception. In returning to the novella, I am thus
not seeking to privilege an originary vision but following the lead of the
opera's creators and public, for whom Carmen bore the indelible trace of
her literary forebear. We shall see, however, that this trace issues in resistance and rewriting as much as in admiration and imitation. This is especially true of the novella's baldly proclaimed misogyny, one of its more
vexed aspects. Somewhat blunted in the opera, it has nonetheless elicited
interpretations ranging from winking complicity to engaged feminist cri-

tique—which is to say, a broad range of performative responses, both on-stage and off.[3] Indeed, the complex performativity of Mérimée's text, as opposed to its mastery of literary codes and conventions, makes of his prose a far richer mine for these various critical and dramatic performances than the competent but rather pedestrian adaptation by Meilhac and Halévy.

Mérimée's stance with regard to his literary output may be character-ized as at once ludic and ironic, and *Carmen* is no exception in this re-spect. He began his literary career with two elaborate hoaxes: A collection of plays, *Le Théâtre de Clara Gazul,* published to great acclaim in 1825, was supposedly the work of a Spanish actress; and a collection of ballads, *La Guzla* (1827), was published as the work of a "Dalmatian bard-cum-outlaw."[4] It is not without interest that in both cases the trav-esty includes a transgender identification, and Mérimée pushed this co-quettishness so far as to have his own engraved portrait, "wearing a mantilla, a necklace and a frilly dress," published as the frontispiece of *Clara Gazul.*[5] Further, both works involve a shift of identity across a northern-southern European cultural divide, with an attendant deploy-ment of *couleur locale,* a hallmark of Mérimée's writing. Although Mérimée was to abandon such obvious pranks, these tendencies may be observed in his later work. With *Carmen* he returned to his favorite for-eign setting, Spain, but bald literary hoax has given way to complex shift-ings of identity and narrative voice and structure through which an array of transgressions is artfully explored.

The first, three-chapter version of *Carmen* was published on October 1, 1845, in the *Revue des Deux Mondes,* a journal that, according to Peter Robinson, "had originally been founded as a bi-weekly travel journal de-picting, for the civilized 'world' of France, exotic landscapes and adven-tures in what today we call the Third World."[6] Robinson is doubtless correct that many readers did not realize that *Carmen* was a fiction rather than another piece of travel writing of the sort published by the *Revue,* and indeed one may say that Mérimée did everything to encourage this reading of his tale as *vérité vécue.* The cultivated readers of the *Revue* would in all likelihood have known that Mérimée had published such accounts of his travels in Spain in other reviews (the *Revue de Paris* and *L'Artiste*) in 1830 and 1831.[7] Especially astute readers might even have noted a com-mon ambiance and recycling of material (bullfights, bandits, etc.) from the earlier letters to the later text. Further, the narrator of the tale is an ama-teur archaeologist who claims to have encountered first Don José, then Carmen, and finally Don José again during his travels in Andalusia while searching for the site of the battle of Munda between Julius Caesar and "the champions of the Republic."[8] He ironically assures those readers

who may be dismayed to see him so quickly abandon his ostensible subject in favor of relating the story of Don José and Carmen that he will
publish a memoir on "le problème géographique qui tient toute l'Europe
savante en suspens" ["the geographic problem that is keeping all of
learned Europe in suspense"] (345). It is a certainty that many if not most
contemporary readers, and later ones as well, took this figure to be
Mérimée himself.[9] Most would have known that Mérimée was something more than an amateur archaeologist, having become Inspector General of Historic Monuments in 1831 and undertaken extensive regional
tours of France to determine the status of historic edifices, especially
those menaced with destruction. And, as an added twist, although
Mérimée did not publish the memoir promised by his narrator, he did
refer to the site of Munda in an article published in 1844 in the *Revue
archéologique*.[10] But *Carmen* is emphatically not a more or less transparent
transposition of Mérimée's travels, and his *alter ego* is more of an artful
mask than a truthful representation of himself, despite the many connections that may exist between them. Here, we see again the way that
Mérimée tended to establish an ironic—and safe—distance between himself and his subject matter. The necessity of these strategies is surely a
complex issue, for this ironic distancing is clearly also between Mérimée
and other aspects of himself. This ironic distance may be construed in psychoanalytic terms as expressing the position of the ego between the
superego and the libidinous urges of the id.[11] I prefer to see in this dynamic the representation (and policing) of the self as other; and in this
representation of the projected other, the theatre in which Mérimée's sexual anxieties are played out. One may surmise that the latter included
more or less suppressed homosexual tendencies for which he overcom-
pensated by a voracious sexual consumption of "othered" women, i.e.,
women of lower social station and/or foreign extraction. At this level,
Carmen may be read as the half-confession of a "sexual tourist"—a role
well documented in Mérimée's private correspondence.

Mérimée was probably the most widely traveled of the romantic authors of his day, especially if one joins to his many trips abroad the yearly
tournées in France that he undertook in his position of inspector general.[12]
In both instances he maintained a voluminous correspondence, both official and private, and in his personal letters he often detailed his sexual
(mis)adventures. The latter tend to take on the characteristics of ethnographic discourse, and they seem less a departure than a continuation in
another domain of his work as inspector-archaeologist. One or two examples will suffice to give the tone of these *confidences*, which express not
so much the excitement of new exploits as they do the ennui of the world-
weary traveler. To a male correspondent he writes:

Figurez-vous que je n'ai pas encore pu trouver de femme. Il y avait bien à Nevers certaines créatures ayant de la gorge, mais elles avaient des bas de laine bleue et des chevilles comme ma cuisse. Je voudrais bien les tenir aujourd'hui. J'ai fait deux lieues dans Autun aujourd'hui sans rien trouver. O Paris, Paris![13]

[Imagine that I have not yet been able to find a woman. Granted, in Nevers there were certain bosomy creatures, but they wore blue wool stockings and had ankles like my thigh. I'd be quite willing to have them today. I covered two leagues in Autun today without finding anything. Oh, Paris, Paris!]

In letters written on the same *tournée,* he comments on the prostitutes of Lyon and Marseille. They are scarce in the former city, he says, because they are persecuted by the king's justice. He observes that, for want of *filles de joie,* the youth of Lyon have recourse to female silk workers, who cost only 40 sous, adding that, in his alarming state, his friend may read that he has been arrested for debauching a gunner (*canonnier*).[14] As for Marseille:

Les putains sont d'une saleté inouïe. Elles ont des bas jaunes d'une epaisseur singulière, jarretière au dessous du genou etc. Toutes ces monstruosités, vous le savez, favorisent l'érection dans certaines circonstances données.[15]

[The whores are extraordinarily filthy. They wear yellow stockings that are remarkably thick, garter above the knee etc. All of these monstrosities, as you know, are conducive to an erection in certain circumstances.]

Mérimée also gave generous advice to male friends embarking on travels to European destinations or more exotic climes. He admonishes his friend the archaeologist Félicien de Saulcy, before the latter's trip to Malta in 1845, that he should always use a condom there and in all of the Levant, adding that if he doesn't have time to seduce the girl at the Hotel Clarence, he can ask the hotel boy for directions to the local whores, who are Phoenician and speak a form of Arabic ("parlant Arabe ou peu s'en faut").[16] In Smyrna, he should ask at the café at the point, past the French consulate, or at a "house" in that quarter. After visiting the whores, he should be sure to see the tomb of Tantalus. Mérimée adds teasingly that he has heard that Saulcy has taken a "glove" along with him, doubtless for abuse of his "little finger." As the last examples indicate, the call of the Orient (which, for the romantics, we have seen, included southern Spain) also held out for European men of a certain social station the possibility of a less inhibited sexuality. Of Flaubert's experience of the Orient, for example, Edward Said remarks on an "almost uniform association between the Orient and sex," noting that this was a "remarkably persistent motif in

Western attitudes to the Orient."[17] Ronald Hyam pithily sums up the question with the statement that "the expansion of Europe was not only a matter of 'Christianity and commerce,' but was also a matter of copulation and concubinage."[18] Mérimée's text cannot of course be reduced to this particular discursive register that we have noted in his correspondence, but neither is his representation of Carmen distorted when seen through this lens. If anything, it brings her into focus.

Mérimée's conception of Carmen may be traced with certainty back to his travels in Spain in 1830. In the fourth of his letters from Spain, "Les sorcières espagnoles," he describes a brief encounter at a country inn with a certain Carmencita, not too dark-complected ("point trop basanée"), who served him water and gazpacho and of whom he then made a portrait in his sketchbook.[19] He would have willingly tarried there longer had his guide Vicente not been extremely impatient to leave. Once again on the road, Vicente remarks that if the gazpacho was so tasty, it is perhaps because it was made by the devil, for the girl's mother is a known sorcerer, and perhaps the girl as well. As for the Carmen story, Mérimée would later claim that it was inspired by an incident related to him by the countess of Montijo, for although he had a predilection for the company of subalterns on his travels in Spain, his encounter with the Montijo family during his 1830 trip would prove to be capital in his future career.[20] In May of 1845 he wrote to the countess:

> Je viens de passer huit jours enfermé à écrire, non point les faits et gestes de feu D. Pedro, mais une histoire que vous m'avez racontée il y a quinze ans, et que je crains d'avoir gâtée. Il s'agissait d'un Jaque de Malaga qui avait tué sa maîtresse, laquelle se consacrait exclusivement au public. . . . Comme j'étudie les bohémiens depuis quelque temps avec beaucoup de soin, j'ai fait mon héroïne bohémienne.[21]

> [I've just spent a week shut in and writing, not the deeds and exploits of the late D[on] Pedro, but a story that you had told me fifteen years ago, and which I'm afraid of having spoiled. It was about a tough from Malaga who had killed his mistress, who devoted herself exclusively to prostitution. . . . Since I've been studying gypsies for some time with great care, I've made my heroine a gypsy.]

He goes on to ask the countess if she is familiar with a book by a certain Mr. Borrow, a translation into gypsy language of the Gospel of Saint Luke. Sorceress, murdered prostitute, gypsy, tantalizing but elusive object of desire: these are the various strands that, filtered through a range of literary and ethnographic sources, would issue in Mérimée's bewitching heroine.

Jean Pommier has argued that Mérimée's travels in Spain were less important in the genesis of *Carmen* than has been generally assumed, but in this regard his fascination with gypsies is especially telling.[22] Mérimée added a fourth chapter to *Carmen*, a sort of ethnographic excursus on the mores and language of gypsies, when he re-published it with two other tales in 1846. For this chapter, he drew on works by the above-mentioned George Borrow (whose Christian zeal made his erudition somewhat suspect to Mérimée) and others, but also on his own research and firsthand encounters with gypsies. Thus in a letter to Ludovic Vitet written in October of 1845, he describes his efforts to see a particular gypsy manuscript:

J'ai pourchassé aux environs de Metz une horde de Bohémiens qui passaient pour posséder un mss. [sic] en Romani historique me disait-on. Je n'ai pas trouvé de mss. mais de fort curieuses gens, ayant admirables figuers. Je leur ai parlé dans le dialecte espagnol, ils m'ont repondu dans le dialecte allemand, et comme dit Epistémon, j'ai failli comprendre.[23]

[I pursued in the area around Metz a horde of Bohemians who were supposed to have a manuscript in Romany (gypsy language), so I was told. I found no manuscripts but some very curious people with admirable faces. I spoke to them in Spanish dialect, they answered me in German dialect, and as Epistemon says, I nearly understood.]

In another letter to the countess of Montijo, written in Barcelona in November of 1846, Mérimée gives a detailed description of a visit to a *tertulia* where gypsies were celebrating the birth of a child.[24] His host in Barcelona was a certain Ferdinand de Lesseps, the French consul and a cousin of the countess, but he also had gypsy friends who would come to visit him, and on this occasion, they invited him to the *tertulia*.[25] He describes the scene thus:

L'événement avait eu lieu depuis deux heures seulement. Nous nous trouvâmes environ trente personnes dans une chambre grande comme celle que j'occupais à Madrid. Il y avait trois guitares et l'on chantait à tue-tête en rommani et en catalan. La société se composait de cinq *gitanas,* dont une assez jolie, et d'autant d'hommes de même race, le reste catalans, voleurs je suppose, ou maquignons, ce qui revient au même.[26]

[The event had taken place only two hours before. There were about thirty of us in a room the size of the one in which I stayed in Madrid. There were three guitars and people were singing at the top of their lungs in Romany and Catalan. The company was made up of five gypsy women, one of them rather pretty, and as many men of the same race, the others being Catalan and thieves, I suppose, or pimps, which amounts to the same thing.]

He tries to give money to an old woman to go buy wine, but one of the gypsy men gives his money back, saying that he is doing too great an honor to his humble abode. He is given wine without paying. Unlike the narrator in *Carmen*, he arrives at home in possession of his money and his watch. Then he observes:

> Les chansons qui m'étaient inintelligibles, avaient le mérite de me rappeler l'Andalousie. On m'en a dicté une en rommani que j'ai comprise. C'est un homme qui parle de sa misère, et qui raconte combien il a été de temps sans manger. Pauvres gens! N'auraient-ils pas été parfaitement justifiables, s'ils m'avaient pris mon argent et mes habits et mis à la porte avec coups de bâton?[27]

> [The songs, which were unintelligible to me, had the merit of recalling Andalusia to my memory. They dictated to me one in Romany, which I understood. It's a man speaking about his misery, who relates how long he has gone without eating. Poor people! Wouldn't they have been perfectly justified if they had taken my money and my clothes and thrown me out after beating me with a stick?]

He concludes this curious account with remarks on the dialect spoken by these gypsies. *Carmen* presents, not surprisingly, the same tensions that we find in this remarkable letter: subjective experience versus scholarly observation, guilty pleasures versus a desire for authenticity—and an undeniable empathy with the object of discourse. These tensions also result in layering and slippages in voice and point of view of the sort that we will see in the novella. Thus, in Mérimée's letter the description of singing gives way to the dictation of a song, and the text invites ambiguity in the way that Mérimée's voice fuses with the words of the song and of the person dictating it to him. We will also find these elements in other, dramatic, performances of *Carmen*.

As noted above, the original version of *Carmen* consisted of three chapters that recounted the narrator's encounters first with Don José, then with Carmen, and finally again with Don José. In chapter three, the narrator becomes the *narrataire* as Don José takes over the narration and relates the story of his tragic affair with Carmen. To these, Mérimée added the fourth chapter, in which the narrator returns to deliver a disquisition on gypsies, thus creating a frame structure around Don José's central narrative, by far the longest of the chapters, and providing closure, a feature that was considerably weaker in the original text. The first two chapters both carry a high erotic charge, and indeed the narrator's twin encounters with his two characters present many similarities. In chapter one, while searching with his guide for the site of the battle of Munda, the narrator becomes thirsty, and having followed a narrow gorge to a spring surrounded by high cliffs, he discovers a man resting there. In chapter two, he meets Carmen in Córdoba by the Guadalquivir, where she has been bathing at dusk with other women from the tobacco factory while he and other men try to spy on them from the bank. In both cases, there is danger lurking in the air. Having heard so much about bandits without ever seeing one, he nonchalantly greets the stranger and drinks from the spring, but he soon realizes that Don José is indeed a bandit. As for Carmen, in a passage that echoes Hugo's above-cited Orientalist construct, in the dark he can't tell if she is Andalusian, Moorish, or Jewish (he says he dares not say the third word aloud), but Carmen quickly puts an end to his uncertainty with a breezy "Allons, allons! vous voyez que je suis bohémienne" ("Come, come! you can see perfectly well that I'm a gypsy") (359). He shares his cigars and then his food with Don José.[28] For Carmen he gallantly puts out his own cigar, then smokes a cigarette with her as their smoke mingles in the night air. The narrator, his guide, and Don José subsequently repair to an isolated inn to spend the night. When he realizes that his guide plans to alert the authorities of the outlaw's whereabouts, he feels bound by the law of hospitality—established by the offering and receiving of the cigar[29]—to inform him of the danger, allowing him to escape certain arrest. As for Carmen, they go first to an ice cream parlor and then to her room, ostensibly so that she can read his fortune. They are interrupted by a furious Don José, who bursts in on them, assuming that Carmen is with another lover. Despite Carmen's gestures urging him to cut the Frenchman's throat (she already has his watch but would like to have his purse as well), when Don José realizes that it is his benefactor, he returns the favor and sends him on his way. Through these many parallels, both Carmen and Don José emerge as dangerous and desirable exotic figures, while Don José also takes on the more ambiguous role of the narrator's double. Fortunately for the narrator, his desire to possess Carmen is never realized and his life is spared. But the man who spares his life, frustrated in his attempts to subdue Carmen, murders her and awaits execution when the narrator meets him again in chapter three.

Perhaps the most intriguing aspect of these initial encounters is the physical descriptions the narrator gives of the two protagonists. The description of Don José shows the tension in the narrator's desire to view him as both brother and other:

C'était un jeune gaillard, de taille moyenne, mais d'apparence robuste, au regard sombre, et fier. Son teint, qui avait pu être beau, était devenu, par l'action du soleil, plus foncé que ses cheveux. (346)

[He was a young fellow, of average height but robust in appearance, with a
look in his eye that was dark, and proud. His complexion, which could once
have been fine, had become, through the action of the sun, darker than his
hair.]

Before learning that this apparent bandit is a Basque of noble descent from
Navarre, he perceives moral and physical qualities in him that neither hard-
ship nor the harshness of the environment have completely effaced. Later,
on their way to the inn, he thinks that his companion matches the picture
he has seen of the notorious bandit José-Maria: "cheveux blonds, yeux
bleus, grande bouche, belles dents, les mains petites" (350) ["blond hair,
blue eyes, large mouth, fine teeth, small hands"]. His new acquaintance is
decidedly not of common Andalusian stock. In fact, he is the very image
of fallen nobility: "sa figure, à la fois noble et farouche, me rappelait le Satan
de Milton" (352) ["his face, at once noble and fierce, reminded me of Milton's Satan"]. It comes as no surprise that he feels a bond with this man
such as he could never have with his local guide and indeed foils the latter's attempt to collect the reward for his capture. When the narrator must
decide which man he will betray, there is no hesitation and only a moment
of self-reproach that he has let himself be swayed by the obligations of hospitality, a "préjugé de sauvage" (356) ["primitive's notion"].

   In Córdoba, as mentioned above, the narrator engages in a favorite local
pastime of the men of that town, trying to catch a glimpse of the women
bathing in the river. Unable to see much of anything, he muses that with
a poetic turn of mind one could fantasize a scene of Diana bathing with
her nymphs without having to fear Acteon's fate. And then, one night, Carmen emerges from the darkness. He first notices the bouquet of jasmine in
her hair, which gives off an intoxicating perfume, then that she is simply,
perhaps even poorly dressed, all in black. And finally, that she is "petite,
jeune, bien faite, et qu'elle avait de très grands yeux" (358) ["short, young,
with a nice body, and that she had very big eyes"]. When he gets her into
the light, a fuller description follows, which is worth quoting at length:

   Je doute que mademoiselle Carmen fût de race pure, du moins elle était in-
   finiment plus jolie que toutes les femmes de sa nation que j'aie jamais ren-
   contrées. Pour qu'une femme soit belle, il faut, disent les Espagnols, qu'elle
   réunisse trente si, ou, si l'on veut, qu'on puisse la définir au moyen de dix
   adjectifs applicables chacun à trois parties de sa personne. Par exemple, elle
   doit avoir trois choses noires: les yeux, les paupières et les sourcils; trois fines,
   les doigts, les lèvres, les cheveux, etc. Voyez Brantôme pour le reste. Ma bohémienne ne pouvait prétendre à tant de perfections. Sa peau, d'ailleurs par-
   faitement unie, approchait de la teinte du cuivre. Ses yeux étaient obliques,
   mais admirablement fendus; ses lèvres un peu fortes, mais bien dessinées et

laissant voir des dents plus blanches que des amandes sans leur peau. Ses cheveux, peut-être un peu gros, étaient noirs, à reflets bleus comme l'aile d'un corbeau, longs et luisants. Pour ne pas vous fatiguer d'une description trop prolixe, je vous dirai en somme qu'à chaque défaut elle réunissait une qualité qui ressortait peut-être plus fortement par le contraste. C'était une beauté étrange et sauvage, une figure qui étonnait d'abord, mais qu'on ne pouvait oublier. Ses yeux surtout avaient une expression à la fois voluptueuse et farouche que je n'ai trouvée depuis à aucun regard humain. Oeil de bohémien, oeil de loup, c'est un dicton espagnol qui dénote une bonne observation. Si vous n'avez pas le temps d'aller au Jardin des Plantes pour étudier le regard d'un loup, considérez votre chat quand il guette un moineau. (360–61)

[I doubt that Miss Carmen was of pure gypsy blood, but in any case she was infinitely more beautiful than all the other women of her race that I have ever met. For a woman to be beautiful, so the Spanish say, she must have thirty "pluses," or, if you will, one must be able to describe her using ten adjectives, each applicable to three parts of her body. For example, she must have three things that are black: her eyes, eyelashes, and eyebrows; three that are delicate, her fingers, lips, and hair, etc. (See Brantôme for the rest.) My gypsy could not lay claim to so many perfections. Her skin, which was moreover perfectly smooth, was close to the color of bronze. Her eyes were slanted but admirably wide; her lips somewhat thick but finely delineated, and allowing a glimpse of teeth whiter than shelled almonds. Her hair, perhaps a bit coarse, was black, with blue highlights like a crow's wing, long and shiny. But so as not to tire you with too wordy a description, I'll sum up by telling you that to each defect she joined a quality that was set off perhaps more strongly by the contrast. Hers was a strange and wild beauty, a face that was disconcerting at first but which one could not forget. Her eyes, especially, had an expression at once sensual and fierce, which I have not seen again in any human gaze. Eye of the gypsy, eye of the wolf, is a Spanish saying that denotes sharp observation. If you don't have time to go to the Botanical Gardens to study a wolf's eyes, just look at your cat when it lies in wait for a sparrow.]

What so strikingly emerges from this passage is the narrator's recognition and alarm when he realizes that conventional canons of beauty, through which the female body is analyzed according to highly codified schemes of classification—the *blason du corps féminin* as it was practiced by authors of Renaissance erotica like Brantôme—utterly fail as a device for representing Carmen. Put another way, this viewpoint positions Carmen outside culture, or at least half in the realm of animal nature. Unlike Don José, who has fallen away from the divine model but still offers a pale reflection of it, Carmen defies it, is set over against it. Such "qualities" as she does possess only serve to underscore her radical alterity. Ultimately, only a say-ing from her own culture can capture her unsettling nature.

In the portrait of Carmen, then, Mérimée presents us with a partial fail-ure of representation, the inability of language to capture alterity. While the tendency in recent criticism has been to read his text as the triumphant expression of northern European colonialist ambitions,[30] I would argue that he exposes his culture's ultimate failure to comprehend, and thus to apprehend, the other onto which it projects its own fantasies and anxieties. His text goes even further, though, in that it allows Carmen, who so per-versely avoids all attempts to possess her, including as an object of lan-guage, to become a *subject* possessed of language and of language's powers of (self-)representation. Carmen, as we have seen, never achieves the status of narrator; she is always represented through the speech of the two narra-tors. Even still, in Don José's narrative especially she always seems to be one step ahead, just out of reach. His descriptions of her are, like the narrator's, failed attempts at portraiture, composed of tantalizing glimpses of an elu-sive object. Thus when he first sees her:

> Elle avait un jupon rouge fort court qui laissait voir des bas de soie blancs avec plus d'un trou, et des souliers mignons de maroquin rouge attachés avec des rubans couleur de feu. Elle écartait sa mantille afin de montrer ses épaules et un gros bouquet de cassie qui sortait de sa chemise. (367)

> [She was wearing a very short red skirt that revealed white silk hose with more than one hole, and dainty red leather shoes tied with fire-colored rib-bons. She drew her mantilla aside so as to show her shoulders and a big bou-quet of acacia flowers that emerged from her blouse.][31]

After the brawl in the factory, when he lets her escape, he is left with a vi-sion of her fleeing legs:

> D'un bond, elle saute par-dessus moi et se met à courir en nous montrant une paire de jambes! . . . On dit jambes de Basque: les siennes en valaient bien d'autres. . . . aussi vites que bien tournées. (372)

> [In a single bound, she jumped over me and started to run, showing us a pair of legs! . . . People talk of "Basque legs": hers were as good as any . . . as fast as they were shapely.]

But it is in her manipulation of language above all that Carmen demon-strates her ability to shift identity, to construct a new persona, to control others. Indeed, her mastery of the performativity of language places Car-men on a par with her two narrators. We see this in her wordplay in her first conversation with the tongue-tied Don José, an encounter that, as Daniel Guichard's analysis shows, has the structure and logic of a duel.[32] Carmen persuades Don José to let her escape by convincing him that she is Basque. Even though she looks nothing like a Basque and butchers the language, her *performance* is utterly convincing. Later, when Don José, disguised as a fruitseller, goes looking for Carmen in Gibraltar and finds her dressed like a queen in the company of an English army officer, she uses Basque to communicate with him and provides "translation" for her companion, who has no idea that he is witnessing a spat between his lady friend and a jealous lover:

—Je donnerais un doigt pour tenir ton mylord dans la montagne, chacun un maquila[33] au poing.
—Maquila, qu'est-ce que ça veut dire? demanda l'Anglais.
—Manquila, dit Carmen riant toujours, c'est une orange. N'est-ce pas un bien drôle de mot pour une orange? Il dit qu'il voudrait vous faire manger du maquila. (391)

["I would give a finger to have your mylord in the mountains, each of us with a *maquila* in his fist."
"What's a *maquila*?" the Englishman asked.
"A *maquila*," said Carmen, still laughing, "is an orange. Isn't that a strange name for an orange? He says he like to make you eat some *maquila*."]

The undertones of this comical scene will soon be realized when Carmen sets up an ambush in which the hapless Englishman is killed by Don José.
     The linguistic games in Mérimée's text are never innocent; conversation and "translation" are duels and sinister endgames. The only match for Carmen on this treacherous playing field is, of course, the narrator, with his mastery of classical languages, his ease of expression in French and Spanish, and his rudimentary knowledge of Basque and Romany. Mérimée tips his hand at the very beginning of his tale, with its misogynistic Greek epigraph from Palladas, which, interestingly enough, he leaves untranslated: "Every woman is bitter as bile, but she has two good moments: in bed and when she's dead." The woman who turns up dead this time is, of course, Carmen, but Mérimée nearly allows her to turn the tables on his narrator, who nervously watches her throat-slitting gestures as she chatters incomprehensibly to Don José in Romany. Peter Robinson argues that, in the end, it is Mérimée who has the last word through his narrator-scholar by ending the ponderous fourth chapter with an enigmatic gypsy proverb that, this time, he translates: "En close bouche, n'entre point de mouche" (409) ["No fly can enter a mouth that is closed"]. Robinson writes: "Writing over Carmen's language, the Frenchman attempts to write her off.

Sealing every possible orifice, the sexual and the verbal, he brings the story to its end—silence."[34] But is it really the Frenchman who has the last word? Is it not the elusive gypsy, whose language wins out in the end in the form of a proverb that the critic Sainte-Beuve saw as a gesture typical of Mérimée:

> It is the equivalent of saying, in salon society, and with that familiar smile: "Of course! Don't be taken in by my brigand and my gypsy any more than you want to be." After having indulged in so much local colour, the author in his turn does not want us to think he is more taken in than is proper.[35]

Rather than subduing his heroine, Mérimée's final gesture seems to be an acknowledgement of the impossibility of doing so, and in this he was characteristically shrewd. Carmen's voice, he must have sensed, would never fall silent.

Manon from the South

When the directors of the Opéra-Comique, Adolphe de Leuven and Camille Du Locle, commissioned a new opera from the young composer Georges Bizet, they were understandably alarmed by the story he chose. Unlike the Opéra, which catered to an upper-class clientele, the Opéra-Comique was a bastion of bourgeois propriety where parents brought their marriageable daughters for interviews with prospective husbands. But Carmen was not the first *femme fatale* or *fille légère* to grace the stages of Paris. Well before the more famous works of Massenet (*Manon*, 1884) and Puccini (*Manon Lescaut*, first version 1893), the abbé Prévost's heroine had been brought to the stage of the Opéra-Comique by two stalwarts of that house's traditions, the composer Auber and his frequent collaborator, the author Eugène Scribe. Even though Bizet wrote *Carmen* against the tradition that Auber and Scribe exemplified, their *Manon Lescaut* of 1856 anticipates Bizet's opera in several respects: the contemporary setting (Scribe updated Prévost's eighteenth-century novel to his day); the main character's sexual infidelities; the undoing of a man of higher social station, whose lapses include desertion from the army; and, most critically, the death of the heroine on stage, a first at the Opéra-Comique.[36] Manon is, in fact, the character to whom Carmen was most often compared by the critics of the day, but unlike Carmen, Manon dies a noble death in the arms of her ever-devoted lover, Des Grieux.[37] Some of Bizet's critics were reminded of Verdi's *La Traviata* (1853; Paris premiere 1864), based on the play *La Dame aux camélias* by Alexandre Dumas *fils*, an author for whom Bizet professed admiration. But Violetta, like Manon, is redeemed in death by her devotion to her lover, Alfredo; indeed, she is nothing short of saintlike in the sacrifice she makes of her own happiness to safeguard his family's respectability. Another heroine may be added to this gallery: Verdi's *Aïda* (1871). Carmen is, if anything, the negative image of "celeste Aïda," but beyond their marked dissimilarities, it is intriguing that both are women of color (Aïda an Ethiopian slave, Carmen a gypsy) in works strongly characterized by a studied deployment of *couleur locale*—a new departure, in fact, in Verdi's operatic output.[38] But Manon, Violetta, and Aïda positively pale next to Carmen and her seemingly endless capacity for transgression, just as Bizet's heroine tends to pale next to her model in Mérimée. And, as we have already suggested, Mérimée's *Carmen* loomed very large indeed over its successor.

Bizet's collaborators on the *Carmen* libretto were two of the most successful writers for the stage at this time, Henri Meilhac and Ludovic Halévy.[39] According to Halévy, Leuven, who was elderly and had also composed works for the Opéra-Comique, was especially horrified at the prospect of seeing Carmen stabbed to death on the stage of the venerable institution, and it fell to Halévy to sell the idea to him. Halévy records Leuven's reaction:

—Carmen! . . . La Carmen de Mérimée! . . . Est-ce qu'elle n'est pas assassinée par son amant? . . . Et ce milieu de voleurs, de bohèmiennes, de cigarières! . . . A l'Opéra-Comique! . . . le théâtre des familles! . . . le théâtre des entrevues de mariages! . . . Nous avons, tous les soirs, cinq ou six loges louées pour ces entrevues . . . Vous allez mettre notre public en fuite . . . c'est impossible![40]

["Carmen! . . . The Carmen of Mérimée! . . . Isn't she murdered by her lover? . . . And this milieu of thieves, gypsies, cigar rollers! . . . At the Comic Opera! . . . The family theater! . . . the theater of marriage interviews! . . . Every evening we rent out five or six boxes for those interviews . . . You are going to drive away our customers . . . it's impossible!]

Leuven implored Halévy to change the ending, but Bizet, whose idea it was to adapt Mérimée's novella to the stage, was adamant that the tragic ending be preserved, as was another major collaborator on the project, Célestine Galli-Marié, the singer who would be the first to play the role (fig. 10.1).[41]

Galli-Marié was not, however, the first singer to whom the role of Carmen was offered. Marie Roze had had this honor, but when she learned that the "scabrous" nature of the role would not be changed, she determined that the role did not suit her, or as she delicately put it, she did not suit the role.[42] It is surprising in retrospect that Galli-Marié was not the obvious first choice. She was brought to the Opéra-Comique in 1862 after

Figure 10.1.   Célestine Galli-Marié. By permission of La Bibliothèque nationale de France, Paris

she was noticed in Rouen by its then director, Emile Perrin, in the title role of an opera by Balfe, *La Bohémienne*. According to Mina Curtiss, as early as 1864 the composer Massé had asked the playwright Victorien Sardou to write a libretto of *Carmen* for her, a project that was never realized.[43] In Paris, she quickly established herself as one of the company's most popular singers, creating the title role in Ambroise Thomas' *Mignon* in 1866. A mezzo appreciated more for her remarkable acting ability than for her singing, she excelled in transvestite roles such as that of Vendredi in Jacques Offenbach's *Robinson Crusoë* and the page Kaled in Aimé Maillart's *Lara*. Of the latter role a critic remarked that it was a "type d'amour sauvage, de passion véhémente, de jalousie africaine, de dévouement assez absolu pour ne pouvoir se changer qu'en haine" ["character of primitive love, of vehement passion, of African jealousy, of devotion absolute enough that it could only become hatred"].[44] In the words of another critic: "La passion jaillit de ses yeux en flèches magnétiques. Ses traits, coupés pour l'expression des grands sentiments héroïques, sont éclairés d'un sourire tout féminin, qui en tempère la robustesse virile" ["Passion shoots from her eyes in magnetic flames. Her features, drawn for the expression of great heroic emotions, are lighted by a most feminine smile, which is tempered by its virile strength"].[45] "Passion" and the ability to perform across a broad spectrum of gender positions were thus the hallmarks of Galli-Marié's art and temperament, and once she had agreed to take on the role of Carmen, she became an ardent champion of Bizet's work and a close partner in its conceptualization. In particular, legend has it that she had Bizet rewrite her entrance thirteen times until they agreed on the Habañera, a setting of a song by the Spanish-Cuban composer Sébastián Yradier for which Bizet himself wrote the words.[46]

The negotiations with Galli-Marié over scheduling, rehearsals, and salary lasted from September to December of 1873. Three days before reaching agreement with Du Locle, she wrote coyly to the "trop sensible directeur" ["too sensitive director"]:

> Qu'allez-vous devenir si j'accepte et si vous me voyez sans les grâces sauvages de la jolie Carmen! Incendie à l'opéra comique alors et nous sommes tout à fait perdus.[47]

> [What will become of you if I accept and if you discover that I don't have the savage charms of the lovely Carmen! They'll burn down the Opéra-Comique, and we will be completely lost.]

She hardly need have worried. By all accounts, even those that were frankly hostile, her performance was remarkable for its "realism" and its dramatic intensity. Surprisingly, she had apparently never read Mérimée's novella before agreeing to play Carmen, but according to Henry Malherbe, she read Mérimée's tale over and over and modeled her performance on the writer's indications.[48] And indeed, the impression of many critics was that Mérimée's heroine had come to life onstage, for better or worse. One reviewer described her performance in these terms:

Si quelqu'un pouvait nous rendre Carmen, telle que l'a rencontrée Prosper Mérimée et telle qu'il l'a chantée, c'est bien certainement Mme Galli-Marié dont la poésie sauvage, la voix chaude, les attitudes provoquantes mettent en relief les moindres détails avec une coleur admirable. Il est impossible d'être plus réaliste et plus poëte à la fois. Comédienne et chanteuse, danseuse même et danseuse fort lascive, Mme Galli-Marié s'est montrée supérieure à tous les points de vue.[49]

[If anyone could render Carmen for us, such as Prosper Mérimée met and celebrated her, it is certainly Mme Galli-Marié, whose savage poetry, warm voice, and provocative attitudes throw the slightest details into relief with admirable color. It is impossible to be more realistic and more poetic at one and the same time. Actress and singer, dancer even, and quite a lascivious dancer at that, Mme Galli-Marié has shown herself to be superlative from all points of view.]

"Bénédict" (B. Jouvin), writing in the *Figaro,* noted however that if Mérimée's heroine had appeared on the stage without Meilhac and Halévy's modifications, the walls of the "virtuous Opéra-Comique" would have fallen![50] For their Carmen, among other changes, is not married to a gypsy when she takes up with Don José, and their path is not littered with the bodies of victims as it is in the novella. There were, of course, reviewers who received the opera with great hostility, and not surprisingly, Galli-Marié in some cases bore the brunt of their wrath. Thus, in an oft-quoted review, Oscar Commettant wrote in the *Revue musicale* that Bizet's "ingenious music," with its "daring use of dissonance" and "instrumental cleverness" was hardly adequate to the representation of "Miss Carmen's uterine ragings" ["les fureurs utérines de Mlle Carmen"] and continued:

L'exécution de *Carmen* aurait, de la part de Mme Galli-Marié, grand besoin d'être amendée. Cette artiste distinguée aurait pu corriger ce que le rôle de la bohémienne, sans coeur, ni foi, ni loi, présentait de choquant et d'antipathique à la scène; elle a, au contraire, exagéré les vices de Carmen par un réalisme qui serait à peine supportable à l'opérette, dans un petit théâtre. A l'Opéra-Comique, théâtre subventionné, théâtre honnête, s'il en fût, Mlle Carmen devrait modérer ses passions.[51]

[The performance of *Carmen*, in so far as Mme Galli-Marié is concerned, could well afford improvement. This distinguished artist could have corrected those aspects of the role of the heartless, faithless, and lawless gypsy that were too shocking and repellent for the stage; she has, on the contrary, exaggerated Carmen's vices by using a realism that would barely be tolerable in an operetta in a small theatre. At the Opéra-Comique, which is a subsidized theatre, a respectable theatre if ever there was one, Miss Carmen should temper her passions.]

Although the tempering of passion was not part of Galli-Marié's conception of the role, several reviewers commented favorably on the subtlety of her dramatic portrayal. She went on to perform *Carmen* to great acclaim in Naples, Brussels, and other European cities before her triumphal return to the Opéra-Comique in her most famous role in 1883. More than any other individual, Galli-Marié was responsible for *Carmen's* success after the death of its composer. She performed the role over 1,200 times, coming out of retirement to sing a final gala performance in Paris on December 11, 1890, to raise funds for a Bizet monument. After her death in 1905, she was buried in the cemetery at Cannes, where she wished to repose near the grave of Prosper Mérimée.

Despite the fierce hostility of some reviewers, it is not true that *Carmen* was met by "almost unrelieved condemnation," as Susan McClary writes, voicing an all too common opinion that *Carmen* was a critical failure at its premiere.[52] The thirty-odd reviews of the premiere that I have examined to date give a much more nuanced impression. Some critics are extremely positive, going so far as to speak of the opera's success; while damning reviews, like Commettant's, are decidedly in the minority. Galli-Marié herself, according to Malherbe, maintained that *Carmen's* "failure" was a legend, an opinion that is seconded in the volume celebrating the centennial of the Opéra-Comique's present theater.[53] Although the opera premiered the evening before the *mi-carême* (Mid-Lent), a particularly festive period during which many theatres, including the Opéra-Comique, hosted costume balls, Malherbe notes that the tense political situation—a government had just fallen and a new one was not yet in place—probably kept many potential theatre-goers at home at a time when memories of the urban insurrection of the Commune were still fresh. When a new government was installed a few days later, *Carmen's* performance at the box office improved noticeably.[54] Malherbe also speculates that the choice of a work on a Spanish theme had been meant to coincide with the reestablishment of diplomatic relations between France and the Spanish government of Alphonso XII, whose forces were locked in a civil war with their Carlist adversaries. President Mac-Mahon received Alphonso's ambassador on the morning of the day of the premiere. Du Locle, nervous about the piece's "immoral" theme, supposedly decided not to invite the visiting Spanish dignitaries and is thought to have published a disclaimer that ran in Parisian papers on the day of the premiere.[55] As Susan McClary remarks, "*Carmen* was always about politics,"[56] and when Bizet's opera had its spectacular premiere at the Paris Opéra in 1959, political and artistic contingencies would once again share center stage.

Of Galli-Marié's successors at the Opéra-Comique, Emma Calvé, the *Carmen* of the one thousandth performance at the Parisian theatre, is beyond any doubt the only singer who rivaled her notoriety in the role (fig. 10.2). Like her illustrious predecessor, Calvé took *Carmen* on the road, performing the opera around the globe. Her fame was such that Queen Victoria had a command performance arranged for her amusement at Windsor Castle in 1891.[57] One particular anecdote indicates the extent to which she became identified with her character. While in New York for performances at the Metropolitan Opera, she went to the post office to pick up a registered letter but had neglected to bring any papers establishing her identity. She received her letter only after giving her rendition of the Habañera for the skeptical postal clerk.[58] Calvé's approach to the role complemented in a sense that of Galli-Marié, who, as we have seen, strove for a maximum of realism and dramatic effect by basing her impersonation on a close reading of Mérimée. Calvé, for her part, sought to achieve a maximum of authenticity. The summer before her *Carmen* debut at the Opéra-Comique in the fall of 1892, she made an expedition to Spain, attending bullfights (the sight of blood made her ill for days) and studying flamenco with gypsy artists in Granada. In her memoirs, she writes:

> Toutes ces danses doivent être de tradition purement arabe, mais agrémentées d'une grâce plus fine, vive et sémillante. . . . Se souvenir surtout qu'une bohémienne n'est jamais commune, ni dans la démarche, ni dans ses gestes, et qu'elle conserve toujours la fierté de sa race, vieille comme le monde![59]

> [All these dances must be from a purely Arabic tradition, but they are enhanced by a gracefulness that is finer, livelier, and more sparkling. . . . Remember above all that a gypsy is never common, neither in her bearing nor in her gestures, and that she always remains proud of her race, which is as old as the world!]

Her debut as Carmen at the Opéra-Comique was a triumph. In the house were Bizet's widow, Geneviève Bizet-Straus, Meilhac and Halévy, and a nostalgic Galli-Marié, who confessed to Calvé that it was the first time that she had ever attended a performance of *Carmen* in which she herself was not performing.[60] Like Galli-Marié's, Calvé's Carmen was a study in dramatic intensity, and her negotiation of Carmen's otherness was achieved through a

Figure 10.2. Emma Calvé. By Permission of La Bibliothèque nationale de France, Paris

similar concern for authenticity that in Calvé's case was nothing short of ethnographic.[61] It is intriguing that, sixty years after Mérimée, Calvé set off to find and study gypsies and, like him, composed a portrait in which her own voice and gestures sought to project an internalized and highly personalized vision of an other at once familiar and strange, inside and outside, shadowy and ostentatious. By the time of Calvé's debut, less than twenty years after Galli-Marié's, this shifting ground had become the rock on which one of France's most cherished cultural symbols was founded.

Cold War Carmen

Shortly after Charles de Gaulle assumed the presidency of the Fifth Republic in 1958, André Malraux, his minister of culture, devised a plan to illustrate the new regime's aggressive cultural program.[62] His idea was for a succession of "grandes premières" at the subsidized theatres of Paris that would, in the words of Daniel-Rops, "faire collaborer activement l'art dramatique au prestige, au rayonnement d'une France qui se veut rénovée" ["to make dramatic art an active collaborator in the prestige and glory of a France determined to show that it has made a new beginning"]. These three gala performances were to be Paul Claudel's *Tête d'or* at the Odéon, now the Théâtre de France, under the direction of Jean-Louis Barrault and Madeleine Renaud; Jean Giraudoux's *Electre* at the Comédie Française; and Bizet's *Carmen* at the Opéra, which was the capstone of the enterprise.[63] The Opéra-Comique had continued to be *Carmen*'s home, of course, where it was performed in the original version that alternated speech and song, the hallmark of the comic opera form. A version with recitatives written by Bizet's fellow composer Ernest Guiraud had been readied for the work's performance in Vienna some six months after the Paris premiere, and it was this version that had met wide acclaim in the other foreign theatres where the opera was performed. No expense was spared in bringing the opera version to the Opéra. It is no exaggeration to say that the third of Malraux's gala evenings, on November 10, 1959, was the cultural event of the year. For it an enormous publicity campaign worthy of a major Hollywood film was put into place. Every aspect of this superproduction received minute attention in the press. The glamorous and the famous turned out in force, doubtless as much to "see and be seen" as to witness what was hyped as an historic event of unequaled importance. Among those on hand, besides De Gaulle and his ministers and a phalanx of civil servants: Jean Cocteau, Coco Chanel, Ingrid Bergman, Marcel Pagnol, Barbara Hutton, René Clair, Jeanne Moreau, and, as one waggish critic put it, a few Rothschild barons. By all accounts, the Shah of Iran's fiancée, Farah Diba, was the most dazzling of those in attendance. Unlike at the Opéra-Comique premiere, the Spanish ambassador was given a seat of honor. As is customary at gala performances when heads of state are in attendance, the Garde Républicaine formed a corridor on the steps leading up to the entrance through which the cultural and governmental elite of France passed. For this occasion, however, the director of the Opéra, A.-M. Julien, added a special touch: in-

Figure 10.3. Jane Rhodes. Photography by Roger Pic. Reproduced by permission of the Bibliothèque Nationale, Paris

terspersed among the Garde were models sent by each of the houses of *haute couture*. The fashions, all in black and white, were created specially for the occasion and the sketches published in the press. Julien had decided too late that the official color of the evening should be "rouge opéra," so to the black-and-white ensembles red accessories were added.

There was also a show on stage. Julien, who was determined to show, as he put it in a press conference, that France had "un magnifique réservoir vocal," cast French singers in the leads.[64] A young, relatively unknown, but rather glamorous soprano, Jane Rhodes (fig. 10.3), sang the title role and gave, by most accounts, a respectable reading, as did the equally glamorous conductor, the former child prodigy Roberto Benzi, later to become Rhodes's husband.

The photogenic Rhodes and Benzi became the darlings of the popular press. Much was made of the former's intellectual and bourgeois pedigree: three certificates in philosophy from the Sorbonne, fencing, classical dance, and tennis with her brothers-in-law, who were champions at the game.[65] As for Benzi, readers were treated to the story of his humble beginnings playing the accordion in the streets of Naples.[66] But the real star of the show was the production itself, which included 350 onstage performers (in addition to the 100 musicians in the pit), fourteen horses, two donkeys, a mule, a dog, and a monkey. What, one may well ask—and people *did* ask—does all of this have to do with *Carmen?* Clearly, as Daniel-Rops saw this, this *Carmen* was a cultural performance in two different but related meanings of the term "culture": an artistic rendering, and a ritualized expression of values shared by a particular group, used as a means of enhancing its power and prestige.

As with *Carmen*'s turbulent premiere at the Opéra-Comique, the Opéra premiere had considerable ramifications of a political nature (in the narrow sense). The production was still running in the spring of 1960 when Nikita Khrushchev came to Paris on a state visit, and de Gaulle took him to see *Carmen* at a second gala performance. Rumors later circulated that it was on this occasion that, during an intermission, de Gaulle informed his guest that France had successfully tested the atomic bomb. Although these rumors were later proven to be false, had de Gaulle wanted to impress Khrushchev with a display of France's might, he could hardly have chosen a better "theatre" in which to do so.[67] Still later, in the fall of 1960, a new singer, Grace Bumbry, was rotated into the cast, making headlines as "la Carmen noire" (in one headline, "la Carmen du Mississipi" [sic]) (fig. 10.4). When the production went on tour to Japan, many French papers ran a photograph that showed "the two Carmens, one white, one black," smiling and waving goodbye at Orly. *Carmen* was clearly taking off for new adventures, adventures that seemed poised less to escape than to continue

Figure 10.4.   Grace Bumbry. By Permission of La Bibliothèque nationale de France, Paris

to play out the historically complex set of attitudes toward the exotic other embodied in the various formulations of Carmen, from Mérimée on.[68]

Notes

1. Victor Hugo, preface to the original edition, *Les Orientales*, ed. Pierre Albouy, *Oeuvres poétiques*, 3 vols. (Paris: Gallimard, 1964-), 1:580. All translations are my own.

2. The *Figaro*'s music columnist, "Un monsieur de l'orchestre" (A. Mortier), relates a humorous incident that supposedly occurred the night of *Carmen*'s premiere. A tiny little dog, completely black, turned up at the door of the theatre director, Camille Du Locle, who decided to take it home after the performance when it developed that no one knew who the owner was. He shut the dog in his office but it managed to escape before the final act. Afterwards, Meilhac is supposed to have remarked to Halévy that it was surely the reincarnation of Mérimée seeking to attend the premiere, a sly reference perhaps to the author's well-known irreligiosity ("Chronique musicale," *Le Figaro*, March 10, 1875).

3. For feminist criticism, see Susan McClary, *Feminist Endings: Music, Gender, and Sexuality* (Minneapolis: University of Minnesota Press, 1991), 56–65; McClary, *Georges Bizet: Carmen*, Cambridge Opera Handbooks (Cambridge: Cambridge University Press, 1992); McClary, "Structures of Identity and Difference in Bizet's *Carmen*," *The Work of Opera: Genre, Nationhood, and Sexual Difference*, ed. Richard Dellamora and Daniel Fischlin (New York: Columbia University Press, 1997), 115–29; Nelly Furman, "The Languages of Love in *Carmen*," *Reading Opera*, ed. Arthur Groos and Roger Parker (Princeton: Princeton University Press, 1988), 168–83; Catherine Clément, *L'Opéra ou la défaite des femmes* (Paris: Grasset, 1979), 94–104.

4. A. W. Raitt, *Prosper Mérimée* (New York: Scribner's, 1970), 37–44.

5. Raitt, *Prosper Mérimée*, 39. See also [Auguste Poulet-Malassis,] *Le Portrait de Prosper Mérimée tour à tour en femme et en homme* (Paris: J. Baur, 1876).

6. Peter Robinson, "Mérimée's *Carmen*," in McClary, *Georges Bizet*, 1.

7. These five "letters from abroad" have been published with four other personal letters in Prosper Mérimée, *Lettres d'Espagne*, ed. Gérard Chaliand (n.p.: Editions Complexe, 1989). Mérimée's offhand remark about the adversaries in the battle of Munda are in fact a veiled expression of his early political sympathies, for, as Chaliand notes, when he made his first trip to Spain, he was a political liberal at a time when known liberal sympathizers were systematically deported or killed under Ferdinand VII (15).

8. Prosper Mérimée, *Carmen*, ed. Maurice Parturier, *Romans et Nouvelles*, 2 vols. (Paris: Garnier, 1967), 2:345, hereafter cited as *Carmen*.

9. Thus, in his review of the Bizet opera, "Savigny" (i.e., Henri Lavoix *père*) wrote: "Si vous vous souvenez bien de ce petit roman, vous vous rappellerez que Mérimée lui-même s'y était mis en scène en se chargeant de rapporter une médaille de José à sa mère, une pauvre femme restée en

peine et ignorante du sort de son fils" ["If you have a good memory of this
little novel, you will recall that Mérimée represented himself in it by taking on the responsibility of bringing back a medallion of José's to his
mother, a poor woman, abandoned in her misery and ignorant of her son's
fate"] (*Illustration*, March 13, 1875). Most reviewers were more astute, but
Ernest Reyer referred cautiously to "le type si admirablement dépeint ou
inventé par Prosper Mérimée" ["the type so admirably depicted or invented by Prosper Mérimée"] (*Journal des Débats*, March 14, 1875).

10. *Carmen*, 659n.

11. For a psychoanalytic interpretation of *Carmen*, see Jacques Chabot, *L'autre
moi: Fantasmes et fantastique dans les Nouvelles de Mérimée* (Aix-en-Provence:
Edisud, 1983):189–212.

12. For Mérimée's foreign travels, see Chaliand's introduction to *Lettres d'Espagne*. For the *tournées* in France, see the editor's introduction to *La naissance des monuments historiques: La correspondance de Prosper Mérimée avec
Ludovic Vitet (1840–1848)*, ed. Maurice Parturier (Paris: Plon, 1934;
CTHS, 1998).

13. Letter from Autun, dated August 15 [1834], to Hippolyte Royer-Collard
(Prosper Merimée, *Correspondance générale*, ed. Maurice Parturier et al., 17
vols. [vols. 1–6, Paris: Le Divan, 1941–52; vols. 7–17, Toulouse: Privat,
1953–64], 1:313), hereafter cited as *CG*.

14. *CG*, 1:316.

15. Ibid., 1:338.

16. Ibid., 4:259–60.

17. Edward Said, *Orientalism* (New York: Vintage, 1979), 188.

18. Ronald Hyam, "Empire and Sexual Opportunity," *Journal of Imperial and
Commonwealth History* 14 (1986): 34–89. The citation appears on page 35.

19. *Lettres d'Espagne*, 108–09.

20. Mérimée's politics shifted steadily to the right over his lifetime. A liberal
when he met the Montijo family, he later became an intimate of their
daughter Eugénie, wife of Napoleon III. The latter appointed him senator
after his marriage to Eugénie.

21. *CG*, 4:295.

22. Pommier points to numerous textual echoes in the tale of works by
Lesage, Prévost, Chateaubriand, Hugo, Musset, and especially Théophile
Gautier, in addition to Mérimée's own writings (Jean Pommier, "Notes sur
Carmen," *Bulletin de la Faculté des lettres de Strasbourg* 8.1 [nov. 1929]: 14–19;
8.2 [déc. 1929]: 51–57; 8.4 [fév. 1930]: 140–45; 8.6 [avril 1930]: 209–16,
esp. 210–16).

23. *CG*, 4:389.

24. In private correspondence, my colleague Salvador Oropesa writes that the
tertulia "used to be the upper hall in old theaters, but in the nineteenth century it was the gathering at the café. Some people met on a regular basis
to talk about art, politics, etc, and they formed sort of closed groups. Some
cafés had shows. All writers had their own *tertulia* or *tertulias*, you could belong to more than one. This is a departure from the culture of the tavern.

Cafés were much more civilized and middle-class places." The context in this passage makes it clear that the *tertulia* in question was a kind of impromptu gypsy cabaret in a private dwelling.

25. It was on this occasion that Mérimée offered to Mme de Lesseps a watercolor he had done of Carmen and Don José, although this identification is perhaps erroneous. The Bibliothèque de l'Opéra possesses what appears to be a colored engraving based on the original watercolor, which is reproduced in Henry Malherbe, *Carmen* (Paris: Albin Michel, 1951).

26. *CG,* 4:559–60.

27. Ibid., 4:560.

28. Robinson brings out the erotic undertones of this scene in what is unfortunately a forced Freudian reading: "When the reader first encounters Don José, he is standing in the earlier-described feminine landscape with a raised blunderbuss. The Frenchman, seeing such endowment, immediately surrenders his cigars" (Robinson, "Mérimée's *Carmen,*" 11).

29. "En Espagne, un cigare donné et reçu établit des relations d'hospitalité, comme en Orient le partage du pain et du sel" (348) ["In Spain, a cigar given and received establishes relations of hospitality, as in the Orient the sharing of bread and salt"].

30. See, for example, David Mickelsen, "Travel, Transgression, and Possession in Mérimée's *Carmen,*" *Romanic Review* 87 (1996): 329–44.

31. We see once again in this description Mérimée's fetishization of feet, ankles, and stockings.

32. Daniel Guichard, "Prosper Mérimée: 'Carmen' (chapitre III)," *L'Ecole des lettres (second cycle)* 15 (15 juin 1986): 25–37; for the analysis in question, 32.

33. The *maquila* is an iron-tipped stick used as a weapon by the Basques. We know that Don José had to leave Navarre because he killed a man in a duel with *maquilas* after a disputed match of *paume.*

34. Robinson, "Mérimée's *Carmen,*" 14. McClary, while she recognizes the complexities of Mérimée's narrative technique, still qualifies it as "monologic" (an odd choice of term), as opposed to the supposed unmediated "polyvocality" of the opera (*Georges Bizet,* 20). Mérimée's text is nothing if not "polyvocalic": Carmen's voice, while certainly mediated in both genres by male narrators or composers and librettists, still manages to come through as distinctively *hers* in both.

35. Sainte-Beuve, *Les grands écrivains français, XIXᵉ siècle: Les romanciers* 2:38, as cited by Raitt, *Prosper Mérimée,* 197.

36. Manon also sings in a cabaret to earn money for Des Grieux, while Carmen's most famous musical number, the Habañera, is a cabaret tune. Carmen also performs, albeit for her own pleasure and for that of her friends, in Lillas Pastia's tavern. See the entry on Auber's *Manon Lescaut* in the *New Grove Dictionary of Opera,* ed. Stanley Sadie, 4 vols. (London: Macmillan, 1994).

37. Sainte-Beuve had remarked of Mérimée's heroine that she was "une *Manon Lescaut* plus poivrée et à l'espagnole" ["a spicier *Manon Lescaut*

served in the Spanish style"] (Sainte-Beuve, *Mes poisons* [Paris: Plasma, 1980], 106). The translation is Raitt's [*Prosper Mérimée*, 195]. The reviewers of Bizet's opera qualified his heroine, or his model, as "une Manon Lescaut bohémienne," "une Manon Lescaut méridionale," "une Manon Lescaut de carrefour," "une Manon sans coeur." One critic, referring to another of Galli-Marié's roles, found her Carmen to be a "Mignon pervertie."

38. It was Camille Du Locle who sent to Verdi the scenario by the archaeologist and Egyptologist Auguste Mariette that would become *Aïda* ("*Aïda*," *New Grove Dictionary of Opera*).

39. Halévy was the cousin of Bizet's wife, Geneviève, who was the daughter of Bizet's teacher, Fromental Halévy, a composer who had composed classic examples of comic opera. For the thousandth performance of *Carmen* in 1905, Ludovic Halévy wrote an article that, although written thirty years after the fact, is one of the most important sources on the premiere ("La millième représentation de Carmen," *Le Théâtre* 145 [janvier 1905]: 5–14). Due to the fact that Bizet had married into a Jewish family, the Nazis sought to ban *Carmen* but, given its tremendous popularity, were ultimately forced to allow its continued staging.

40. Halévy, "La millième représentation," 6.

41. McClary, *Georges Bizet,* 19 (on Bizet's choice of the subject) and 24 (on Galli-Marié's role). Leuven's resignation as director early in 1874 was, according to McClary, due in part to his dissatisfaction over the *Carmen* project (23).

42. Rémy Stricker, *Georges Bizet* (Paris: Gallimard, 1999), 87; Mina Curtiss, *Bizet and His Work* (New York: Knopf, 1958), 355.

43. Curtiss, *Bizet,* 357.

44. Albert Vizentini, "Les jeunes premières du jour," *L'Eclair* (December 8, 1867).

45. Paul Mahalin, *Les Jolies Actrices de Paris,* first series (Paris: Tresse, 1868–78), 31.

46. McClary, *Georges Bizet,* 26; Stricker, 217. Stricker prints the Yradier song, "El Areglito," in an appendix (311–14).

47. Bibliothèque de l'Opéra, dossier Galli-Marié, letter dated December 15, 1873.

48. Malherbe, *Carmen,* 198.

49. Unidentified review in press file on *Carmen* (Bibliothèque nationale de France, nouv. acq. fr., 14352), piece 2. Note the already-mentioned conflation of Mérimée with his narrator.

50. *Le Figaro,* March 5, 1875.

51. *Revue musicale,* March 8, 1875.

52. McClary, *Georges Bizet,* 113. Among the other legends that have become attached to *Carmen:* that Galli-Marié and Bizet were lovers; that, due to the failure of the opera and the breaking off of their relationship, Bizet became deeply depressed; that Bizet committed suicide out of despair (Malherbe, *Carmen,* 91–98). None of these legends has any credence.

Moreover, the opera was by no means a failure in all eyes. Among the more favorable reviews, in addition to the oft-cited one by Théodore de Banville (*National*, March 8, 1875), are the reviews by Armand Gouzien (*Evénement*, March 5, 1875), Pierre Véron (*Charivari*, March 6, 1875), Victorin Joncières (*Liberté*, March 8, 1875), and Ernest Reyer (*Journal des débats*, March 14, 1875).

53. Malherbe, *Carmen*, 92; Michel Parouty, *L'Opéra Comique* (Paris: ASA, 1998), 77–78.
54. Malherbe, *Carmen*, 189. The critic Jules Guillemot noted in his review that not since 1873 had the Parisian public been able to enjoy the creation of a new full-length work at the Opéra-Comique and that on that occasion they had entered the theatre under Thiers's government and left under Mac-Mahon's (*Le Soleil*, March 6, 1875).
55. McClary, *Georges Bizet*, 23; Malherbe, *Carmen*, 190.
56. McClary, *Georges Bizet*, 124.
57. Emma Calvé, *Sous tous les ciels j'ai chanté* (Paris: Plon, 1940), 90–91. The performance was well received. Afterwards, Calvé was rewarded with a portrait of Victoria with the dedication "To Emma Calvé, the beautiful inspired artist. Victoria, Queen-Empress."
58. Calvé, *Sous tous les ciels*, 115.
59. Ibid., 78.
60. Ibid., 84.
61. George Bernard Shaw confessed that Calvé's performance in *Carmen* so overpowered him that he could not properly exercise his job as music critic. See the review of an 1894 performance at Covent Garden, reprinted in *Shaw's Music: The Complete Musical Criticism of Bernard Shaw*, ed. Dan H. Lawrence, 3 vols., second revised edition (London: Bodley Head, 1989), 3:224.
62. This account is based on the files of press clippings for the 1959–60 production: Bibliothèque de l'Opéra, *Carmen*, dossier d'oeuvre, R1-R2 1959.
63. Daniel-Rops, "Générales parisiennes," *L'Echo Liberté de Lyon*, April 11, 1959.
64. As quoted in *Combat*, October 30, 1959.
65. *Paris Match*, November 14, 1959.
66. Ibid., November 9, 1959.
67. I would like to thank M. Philippe Cousin, bibliothécaire adjoint, Bibliothèque de l'Opéra, for sharing this information with me. According to M. Cousin, researchers from the Centre d'Etudes Nucléaires came to the library to ascertain whether or not the rumors were true and found that the dates of the bomb test were after the date of the gala.
68. I would like to thank the staff of the Bibliothèque de l'Opéra and its head curator, M. Pierre Vidal, for the gracious assistance they provided me as I carried out the research for this essay.

Contributors

KATHLEEN M. ASHLEY is Professor of English at the University of Southern Maine. She is the author or editor of *Writing Faith: Text, Sign, and History in the Miracles of Sainte Foy*, with Pamela Sheingorn (Chicago 1999), *Autobiography and Postmodernism*, with Leigh Gilmore and Gerald Peters (Massachusetts 1994), *Interpreting Cultural Symbols: Saint Anne in Late Medieval Society*, with Pamela Sheingorn (Georgia 1990), and *Victor Turner and the Construction of Cultural Criticism* (Indiana 1990), and has published numerous essays on drama.

MARVIN CARLSON is the Sidney E. Cohn Distinguished Professor of Theatre and Comparative Literature at the Graduate Center of the City University of New York. He is the author of numerous books on the history and theory of Western theatre and performance, a recipient of the George Jean Nathan Award in dramatic criticism, and founding editor of the journal *Western European Stages*.

XIAOMEI CHEN is Associate Professor of Asian Literatures at the Ohio State University. She is the author of *Occidentalism: A Theory of Counter Discourse in Post-Mao China* (Oxford 1995) and articles on Chinese theatre in such journals as *New Literary History*, *Representations*, *Critical Inquiry*, the *Journal of Asian Studies*, and *Comparative Literary Studies*.

ROBERT L. A. CLARK is Associate Professor of Romance Languages, Kansas State University. Along with Kathleen Ashley, he is co-editor of *Medieval Conduct: Theories, Histories, Texts* (forthcoming from Minnesota). He has published articles on medieval drama, subjectivity, and cross-dressing in *New Literary History* and the *Journal of Medieval and Early Modern Studies*, among others.

CLAUS CLÜVER is Professor Emeritus of Comparative Literature at Indiana University. He is the author of *Thornton Wilder und André Obey* (Bouvier 1978) and numerous essays. His research and teaching, in the United States, Germany, and Brazil, focus on cross-media studies, especially avant-garde poetics and performance.

CYNTHIA ERB is Associate Professor of English and Co-Director of the Film Studies program at Wayne State University in Detroit. Her publications include *Tracking King Kong: A Hollywood Icon in World Culture* (Wayne State University Press, 1998) as well as articles and reviews in *Cinema Journal*, *Journal of Film and Video*, and *Film*

Quarterly. She is currently working on a book-length study of New Age Hollywood and American religions.

JINHEE KIM is Assistant Professor in the Department of East Asian Languages and Cultures at the University of Southern California, Los Angeles. She received her doctoral degree from Indiana University, where she wrote her dissertation, "Disembodying the Other: East-West Relations and Modern Korean Drama." Before coming to USC, she taught at Smith College. With a grant from the Korea Culture and Arts Foundation, she is preparing a manuscript for *The Anthology of Modern Korean Drama.*

SHELDON H. LU is Associate Professor of Chinese Literature and Film Studies at the University of Pittsburgh. He is the author of *From Historicity to Fictionality: The Chinese Poetics of Narrative* (Stanford University Press, 1994) and editor of *Transnational Chinese Cinemas: Identity, Nationhood, Gender* (University of Hawaii Press, 1997). He is completing a new book: *China, Transnational Visuality, Global Postmodernity.* His critical essays have appeared in many journals.

CLAIRE SPONSLER is Associate Professor of English at the University of Iowa. She is the author of *Drama and Resistance: Bodies, Goods, and Theatricality in Late Medieval England* (Minnesota 1997), as well as numerous articles on medieval culture and performance that have appeared in *New Literary History,* the *Journal of Medieval and Early Modern Studies, Theatre Annual, Theatre Survey,* and other journals and collections. She is completing a book on appropriations of medieval drama in America.

KAREN A. WINSTEAD is Associate Professor of English at the Ohio State University. A specialist in medieval hagiography, she is the author of *Virgin Martyrs: Legends of Sainthood in Late Medieval England,* editor of John Capgrave's *Life of Saint Katherine of Alexandria,* and editor/translator of *Chaste Passions: Medieval English Virgin Martyr Legends.* She is currently working on studies of the fifteenth-century writer John Capgrave and of the Katherine of Alexandria legend.

Select Bibliography

Abrahams, Roger. *The Man-of-Words in the West Indies: Performance and the Emergence of Creole Culture.* Baltimore: Johns Hopkins University Press, 1983.

Anglo, Sydney. *Spectacle, Pageantry, and Early Tudor Policy.* Oxford: Clarendon Press, 1969.

Appadurai, Arjun. "Putting Hierarchy in Its Place." *Cultural Anthropology* 3 (1988): 36–48.

Arróniz, Othón. *Teatro de evangelización en Nueva España.* Mexico: Universidad Nacional Autónoma de México, 1979.

Ashcroft, Bill, Gareth Griffiths, and Helen Tiffin. *Key Concepts in Post-Colonial Studies.* London: Routledge, 1998.

Ashley, Kathleen and Pamela Sheingorn. *Writing Faith: Text, Sign, and History in the Miracles of Sainte Foy.* Chicago: University of Chicago Press, 1999.

Balme, Christopher B. *Decolonizing the Stage.* Oxford: Clarendon Press, 1999.

Bann, Stephen, ed. *Concrete Poetry: An International Anthology.* London: London Magazine Editions, 1967.

Banu, Georges. "Mei Lanfang: A Case Against and a Model for the Occidental Stage." Trans. Ella L. Wiswell and June V. Gibson. *Asian Theatre Journal* 3 (1986): 153–78.

Bhabha, Homi K. *The Location of Culture.* London: Routledge, 1994.

Bishop, T. G. *Shakespeare and the Theatre of Wonder.* Cambridge: Cambridge University Press, 1996.

Bitterli, Urs. *Cultures in Conflict: Encounters Between European and Non-European Cultures, 1492–1800.* Trans. Ritchie Robertson. Stanford: Stanford University Press, 1989.

Blakemore Evans, M. *The Passion Play at Lucerne.* New York: MLA, 1943.

Bloom, Harry and Pat Williams. *King Kong: An African Jazz Opera.* London: Collins, 1961.

Borgatti, Jean M. and Richard Brilliant. *Likeness and Beyond: Portraits from Africa and the World.* New York: The Center for African Art, 1990.

British Concrete Poetry. Ed. John Sharkey. London: Lorrimer, 1971.

Brown, Peter. *Society and the Holy in Late Antiquity.* Berkeley and Los Angeles: University of California Press, 1982.

Bryant, Lawrence. *The King and the City in the Parisian Royal Entry Ceremony: Politics, Ritual, and Art in the Renaissance.* Geneva: Droz, 1986.

Bulatov, Dmitry. *A Point of View: Visual Poetry: The 90s, An Anthology* (Königsberg: Simplisii, 1998).

Burns, Bradford E. *A Documentary History of Brazil.* New York: Knopf, 1966.

A Calendar of the Dramatic Records in the Books of the Livery Companies of London, 1485–1640. Ed. Jean Robertson and Donald Gordon. Malone Society Collections 3. London: Oxford University Press, 1954.

Calvé, Emma. *Sous tous les ciels j'ai chanté.* Paris: Plon, 1940.

Cameron, Kenneth M. *Africa on Film: Beyond Black and White.* New York: Continuum, 1994.

C'est la deduction du Somptueux ordre, plaisantz spectacles et magnifiques theatres dresses et exhibes, par les citoiens de Rouen. . . . Rouen: Robert et Jehan dictz Dugord, 1551.

Chabot, Jacques. *L'autre moi: Fantasmes et fantastique dans les Nouvelles de Mérimée.* Aix-en-Provence: Edisud, 1983.

Chen, Xiaomei. *Occidentalism.* New York: Oxford University Press, 1995.

Choi, Chungmoo. "The Discourse of Decolonization and Popular Memory: South Korea." *Positions* 1 (1993): 77–102.

Ci, Jiwei. *Dialectic of the Chinese Revolution.* Stanford: Stanford University Press, 1994.

Ciecko, Anne T. and Sheldon H. Lu. "The Heroic Trio: Anita Mui, Maggie Cheung, and Michele Yeoh—Self-Reflexivity and the Globalization of the Hong Kong Action Heroine." *Post Script* 19 (1999): 70–86.

Clark, Robert L. A. and Claire Sponsler. "Othered Bodies: Racial Cross-Dressing in the *Mistère de la Sainte Hostie* and the Croxton *Play of the Sacrament.*" *Journal of Medieval and Early Modern Studies* 29 (1999): 61–87.

Clément, Catherine. *L'Opéra ou la défaite des femmes.* Paris: Grasset, 1979.

Clifford, James. *Routes: Travel and Translation in the Late Twentieth Century.* Cambridge: Harvard University Press, 1997.

Clüver, Claus. "Concrete Poetry Into Music: Oliveira's Intersemiotic Transposition." *The Comparatist* 6 (1982): 3–15.

———. "*Klangfarbenmelodie* in Polychromatic Poems: A. von Webern and A. de Campos." *Comparative Literature Studies* 18 (1981): 386–98.

———. "Traduzindo Poesia Visual." *Cânones & Contextos* (Rio de Janeiro: ABRALIC, 1997). 311–27.

Cohen, Gustave. *Histoire de la Mise en Scène dans le Théâtre Religieux français du Moyen Age.* Paris: Champion, 1951.

Coplan, David. *In Township Tonight!: South Africa's Black City Music and Theatre.* London: Longman, 1985.

Curtiss, Mina. *Bizet and His Work.* New York: Knopf, 1958.

Daston, Lorraine and Katharine Park. *Wonders and the Order of Nature, 1150–1750.* New York: Zone Books, 1998.

de Certeau, Michel. *The Practice of Everyday Life.* Trans. Steve Rendall. Berkeley and Los Angeles: University of California Press, 1988.

Delehaye, Hippolyte. *Les Saints Stylites.* Brussels: Société des Bollandistes, 1923.

Eco, Umberto. *The Role of the Reader: Explorations in the Semiotics of Texts.* Bloomington: Indiana University Press, 1979.

Les Entrées royales françaises de 1328 à 1515. Ed. Bernard Guenée and Françoise Lehouz. Paris: Éditions du Centre National de la Recherche Scientifique, 1968.

Erb, Cynthia. *Tracking King Kong: A Hollywood Icon in World Culture.* Detroit: Wayne State University Press, 1998.

European Theories of the Drama. Ed. Barrett H. Clark. New York: Crown, 1965.

Experimental—Visual—Concrete: Avant-Garde Poetry Since the 1960s. Ed. K. David Jackson, Eric Vos, and Johanna Drucker. Amsterdam: Rodopi, 1996.

Fabian, Johannes. *Time and the Other: How Anthropology Makes Its Object.* New York: Columbia University Press, 1983.

Farmer, Sharon. *Communities of Saint Martin: Legend and Ritual in Medieval Tours.* Ithaca: Cornell University Press, 1991.

Feest, Christian. *Indians and Europe: An Interdisciplinary Collection of Essays.* Aachen: Rader Verlag, 1987.

Flanigan, C. Clifford. "Liminality, Carnival and Social Structure: The Case of Late Medieval Biblical Drama." *Victor Turner and the Construction of Cultural Criticism: Between Literature and Anthropology.* Ed. Kathleen M. Ashley. Bloomington: Indiana University Press, 1990.

Foreman, Carolyn Thomas. *Indians Abroad, 1493–1938.* Norman: University of Oklahoma Press, 1943.

Foster, Stephen William. "The Exotic as a Symbolic System." *Dialectical Anthropology* 7 (1982): 21–30.

Furman, Nelly. "The Languages of Love in *Carmen.*" *Reading Opera.* Ed. Arthur Groos and Roger Parker. Princeton: Princeton University Press, 1988. 168–83.

Glasser, Mona. *King Kong: A Venture in the Theatre.* Cape Town: Norman Howell, 1960.

Glissant, Edouard. *Le Discours Antillais.* Paris: Seuil, 1981.

Global/Local: Cultural Production and the Transnational Imaginary. Ed. Rob Wilson and Wimal Dissanayake. Durham: Duke University Press, 1996.

Godzich, Wlad. "Foreword: The Further Possibility of Knowledge." Michel de Certeau. *Heterologies: Discourse on the Other.* Trans. Brian Massumi. Minneapolis: University of Minnesota Press, 1986.

Grabar, Andre. *Early Christian Art: From the Rise of Christianity to the Death of Theodosius.* New York: Odyssey Press, 1968.

Greenblatt, Stephen. *Marvellous Possessions: The Wonder of the New World.* Chicago: University of Chicago Press, 1991.

Greene, David Mason. "The Welsh Characters in Patient Grissil." *Boston University Studies in English* 4 (1960): 171- 80.

Greene, Jody. "New Historicism and Its New World Discoveries." *Yale Journal of Criticism* 4 (1991): 163–98.

Gregory of Tours. *The History of the Franks.* Trans. Lewis Thorpe. New York: Penguin, 1974.

———. *Gregorii Episcopi Turonensis Historiarum Libri X.* Ed. Bruno Krusch. Monumenta Germaniae Historica: Scriptores Rerum Merovingicarum, Vol. 1. Hanover, 1885.

Halévy, Ludovic. "La millième représentation de Carmen." *Le Théâtre* 145 (1905): 5–14.

Hall, Edith. *Inventing the Barbarian: Greek Self-Definition Through Tragedy.* Oxford: Clarendon Press, 1989.

Harris, Max. *The Dialogical Theatre: Dramatizations of the Conquest of Mexico and the Question of the Other.* New York: St. Martin's, 1993.

Hemming, John. *Red Gold: The Conquest of the Brazilian Indians, 1500–1760.* Cambridge: Harvard University Press, 1978.

Hockett, Charles F. *A Course in Modern Linguistics.* New York: Macmillan, 1958.

Holub, Robert C. *Reception Theory: A Critical Introduction.* London: Methuen, 1984.

Hugo, Victor. *Les Orientales.* Ed. Pierre Albouy. *Oeuvres poétiques.* 3 vols. Paris: Gallimard, 1964-.

Hyam, Ronald. "Empire and Sexual Opportunity." *Journal of Imperial and Commonwealth History* 14 (1986): 34–89.

Imaged Words & Worded Images. Ed. Richard Kostelanetz. New York: Outerbridge & Dienstfrey/Dutton, 1970.

The Intercultural Performance Reader. Ed. Patrice Pavis. London: Routledge 1996.

Jameson, Fredric. "Postmodernism, or the Cultural Logic of Late Capitalism." *New Left Review* 46 (1984): 53–92.

———. "Third-World Literature in the Era of Multinational Capitalism." *Social Text* 15,3 (1986): 65–88.

Jauss, Hans Robert. *Toward an Aesthetic of Reception.* Trans. Timothy Bahti. Minneapolis: University of Minnesota Press, 1982.

Kaplan, Paul. *The Rise of the Black Magus in Western Art.* Ann Arbor: UMI Research Press, 1987.

Kavanagh, Robert Mshengu. *Theatre and Cultural Struggle in South Africa.* London: Zed, 1985.

Kipling, Gordon. *Enter the King: Theatre, Liturgy, and Ritual in the Medieval Civic Triumph.* New York: Clarendon, 1998.

Konkrete Poesie: Deutschsprachige Autoren. Ed. Eugen Gomringer. Stuttgart: Philipp Reclam, 1972.

Lea, K. M. *Italian Popular Comedy.* New York: Russell and Russell, 1962.

Lowe, Lisa. *Critical Terrains: French and British Orientalisms.* Ithaca: Cornell University Press, 1991.

Lu, Sheldon H. "Filming Diaspora and Identity: Hong Kong and 1997." *The Cinema of Hong Kong: History, Form, Genre.* Ed. David Desser and Poshek Fu. London and New York: Cambridge University Press, 2000.

———. "Universality/Difference: The Discourses of Chinese Modernity, Postmodernity, and Postcoloniality." *Journal of Asian Pacific Communication* 9 (1999): 97–111.

Malherbe, Henry. *Carmen.* Paris: Albin Michel, 1951.

Mason, Peter. *Deconstructing America: Representations of the Other.* London: Routledge, 1990.

McClary, Susan. *Feminist Endings: Music, Gender, and Sexuality.* Minneapolis: University of Minnesota Press, 1991.

———. *Georges Bizet: Carmen.* Cambridge Opera Handbooks. Cambridge: Cambridge University Press, 1992.

———. "Structures of Identity and Difference in Bizet's *Carmen.*" *The Work of Opera: Genre, Nationhood, and Sexual Difference.* Ed. Richard Dellamora and Daniel Fischlin. New York: Columbia University Press, 1997. 115–29.

McCullough, Kathleen. *Concrete Poetry: An Annotated International Bibliography, With an Index of Poets and Poems.* Troy, NY: Whitston Publishing, 1989.

McGowan, Margaret M. "Forms and Themes in Henri II's Entry into Rouen." *Renaissance Drama* 1 (1968): 199–252.

McIntock, Anne. *Imperial Leather.* New York: Routledge, 1995.

Medieval Drama. Ed. David Bevington. Boston: Houghton Mifflin, 1975.

Mellinkoff, Ruth. *Outcasts: Signs of Otherness in Northern European Art of the Late Middle Ages.* Berkeley and Los Angeles: University of California Press, 1993.

Meng, Yue. "Female Images and National Myth." *Gender Politics in Modern China.* Ed. Tani E. Barlow. Durham: Duke University Press, 1993. 118–136.

Mérimée, Prosper. *Carmen.* Ed. Maurice Parturier. *Romans et Nouvelles.* 2 vols. Paris: Garnier, 1967.

———. *Lettres d'Espagne.* Ed. Gérard Chaliand. N.p.: Editions Complexe, 1989.

Mickelsen, David. "Travel, Transgression, and Possession in Mérimée's *Carmen.*" *Romanic Review* 87 (1996): 329–44.

Modisane, Bloke. *Blame Me on History.* 1963; reprint, London: Penguin, 1990.

Morison, Samuel Eliot. *The European Discovery of America.* 2 vols. New York: Oxford University Press, 1971–74.

Mullaney, Steven. *The Place of the Stage: License, Play, and Power in Renaissance England.* Chicago: University of Chicago Press, 1988.

La Naissance des monuments historiques: La correspondance de Prosper Mérimée avec Ludovic Vitet (1840–1848). Ed. Maurice Parturier. Paris: Plon, 1934; CTHS, 1998.

Nakasa, Nathaniel. *The World of Nat Nakasa.* Ed. Essop Patel. Johannesburg: Ravan, 1985.

Newton, Stella Mary. *Renaissance Theater Costume and the Sense of the Historic Past.* London: Rapp and Whiting, 1975.

Nixon, Rob. *Homelands, Harlem, and Hollywood: South African Culture and the World Beyond.* New York: Routledge, 1994.

Paik, Nak-chung. "The Idea of Korean National Literature Then and Now." *Positions* 1 (1993): 553–580.

Parouty, Michel. *L'Opéra Comique.* Paris: ASA, 1998.

Pavis, Patrice. *Languages of the Stage.* New York: Performing Arts Journal Publications, 1982.

Platt, Peter G. *Reason Diminished: Shakespeare and the Marvelous.* Lincoln: University of Nebraska Press, 1997.

The Play Out of Context: Transferring Plays from Culture to Culture. Ed. Hanna Scolnicov and Peter Holland. Cambridge and New York: Cambridge University Press, 1989.

Poesia Concreta em Portugal. Ed. José Alberto Marques and E. M. de Melo e Castro. Lisboa: Assírio & Alvim, 1973.

Political Shakespeare: New Essays in Cultural Materialism. Ed. Jonathan Dollimore and Alan Sinfield. Manchester: Manchester University Press, 1985.

Poster, Mark. "The Question of Agency: Michel de Certeau and the History of Consumerism." *Diacritics* 22 (1992): 94–107.

Potter, Robert. "Abraham and Human Sacrifice: The Exfoliation of Medieval Drama in Aztec Mexico." *New Theatre Quarterly* 8 (1986): 306–12.

Pratt, Mary Louise. *Imperial Eyes: Travel Writing and Transculturation*. New York: Routledge, 1992.

"Race," Writing, and Difference. Ed. Henry Louis Gates, Jr. Chicago: University of Chicago Press, 1986.

Raitt, A. W. *Prosper Mérimée*. New York: Scribner's, 1970.

Ravicz, Marilyn Ekdahl. *Early Colonial Religious Drama in Mexico*. Washington, D.C.: Catholic University of America Press, 1970.

Richards, David. *Masks of Difference: Cultural Representations in Literature, Anthropology and Art*. Cambridge: Cambridge University Press, 1994.

Robertson, Roland. *Globalization: Social Theory and Global Culture*. London: Sage Publications, 1992.

Ross, Robert. *A Concise History of South Africa*. Cambridge: Cambridge University Press, 1999.

Rouillard, Clarence D. *The Turk in French History, Thought and Literature (1520–1660)*. Paris: Boivin, 1941.

Said, Edward. *Orientalism*. 1978; repr. New York: Random House, 1985.

Shaw's Music: The Complete Musical Criticism of Bernard Shaw. Ed. Dan H. Lawrence. 3 vols. 2nd revised ed. London: Bodley Head, 1989.

Shram, Stuart R. *The Thoughts of Chairman Mao Tse-Tung*. London: Library 33 Limited, 1967.

Solt, John. *Shredding the Tapestry of Meaning: The Poetry and Poetics of Kitasono Katue (1902–1978)*. Cambridge: Harvard University Press, 1999.

Solt, Mary Ellen, ed. *Concrete Poetry: A World View*. Bloomington: Indiana University Press, 1970.

Sponsler, Claire. "The Culture of the Spectator: Conformity and Resistance in Medieval Drama." *Theatre Journal* 44 (1992): 15–29.

———. "Medieval America: Drama and Community in the English Colonies, 1580–1610." *Journal of Medieval and Early Modern Studies* 28 (1998): 453–78.

The Staging of Religious Drama in Europe in the Later Middle Ages: Texts and Documents in English Translation. Ed. Peter Meredith and John E. Tailby. EDAM Monograph Series 4. Kalamazoo: Medieval Institute Publications, 1983.

States, Bert O. *Great Reckonings in Little Rooms: On the Phenomenology of Theater*. Berkeley and Los Angeles: University of California Press, 1985.

Stewart, Susan. *On Longing: Narratives of the Miniature, the Gigantic, the Souvenir, the Collection*. Baltimore: Johns Hopkins University Press, 1984.

Stricker, Rémy. *Georges Bizet*. Paris: Gallimard, 1999.

Strong, Roy. *Art and Power: Renaissance Festivals 1450–1650*. Woodbridge, Suffolk: Boydell Press, 1984.

Teoria da Poesia Concreta: Textos Críticos e Manifestos 1950–1960. Eds. Augusto de Campos, Décio Pignatari, and Haroldo de Campos. 3rd. ed. São Paulo: Brasiliense, 1987.

The Theatre of Medieval Europe: New Research in Early Drama. Ed. Eckhard Simon. Cambridge: Cambridge University Press, 1991.

Trexler, Richard. *Bearing Gifts: The Magi Cult and the Documentation of Social Processes.* Binghamton: Fernand Braudel Center, 1980.

Vecellio, Cesare. *Renaissance Costume Book.* New York: Dover Publications, 1977.

Victor Turner and the Construction of Cultural Criticism: Between Literature and Anthropology. Ed. Kathleen M. Ashley. Bloomington: Indiana University Press, 1990.

Wasserman, Renata R. Mautner. *Exotic Nations: Literature and Cultural Identity in the United States and Brazil, 1830–1930.* Ithaca: Cornell University Press, 1994.

Webster, Michael. *Reading Visual Poetry after Futurism: Marinetti, Apollinaire, Schwitters, Cummings.* New York: Peter Lang, 1995.

Wheaton, Barbara Ketcham. *Savouring the Past: The French Kitchen and Table from 1300 to 1789.* London: Chatto & Windus, 1983.

White-Haired Girl. Trans. Gladys Yang and Yang Hsien-yi. Beijing: Beijing waiwen chubanshe, 1954.

Wickham, Glynne. *Early English Stages, 1300–1660.* 3 vols. in 4. London: Routledge and Kegan Paul, 1981.

Williams, Emmett, ed. *An Anthology of Concrete Poetry.* New York: Something Else Press, 1967.

Winstead, Karen A. "The Transformation of the Miracle Story in the *Libri Historiarum* of Gregory of Tours." *Medium Aevum* 59 (1990): 1–15.

Wintroub, Michael. "Civilizing the Savage and Making a King: The Royal Entry Festival of Henri II (Rouen, 1550)." *Sixteenth Century Journal* 29 (1998): 465–94.

Yi, Man Hûi. *Pul chom kkô chuseyo [Please Turn out the Lights].* Seoul: Taehakro Kûkchang, 1992.

Index